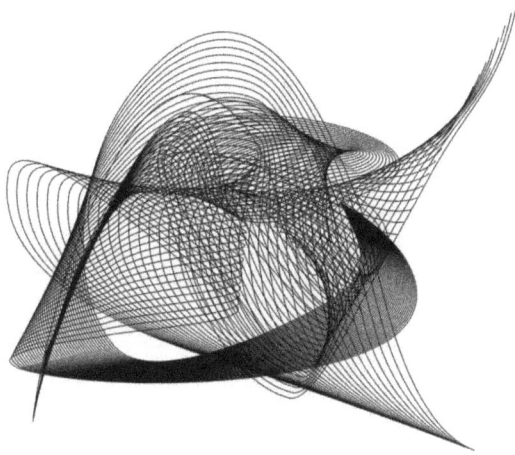

The Agile Brand, Revisited

Principles for the continuously improving,

customer-focused enterprise |

2025 Edition

By Greg Kihlström

Published by:

Agile Brand, LLC
3100 Clarendon Boulevard #200
Arlington, VA 22201

https://www.gregkihlstrom.com

First Edition: January 2025

The publisher is not responsible for websites (or their content) that are not owned by the publisher.

Edited by Loretha Greene, Janelle Kihlström, and Anne-Marie Montague

Cover design and illustrations by Alicia Recco and Greg Kihlström

ISBN: 979-8-9922806-0-9

Contents

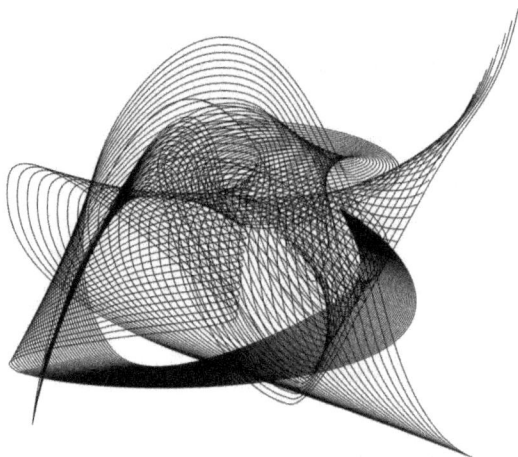

Foreword

Hello. It's nice to meet you. My name is Chris Bach. I figured if you are going to read a foreword with an opinion on our current digital landscape the least the writer can do upfront is warn you of their background.

I'm an entrepreneur who today invests in and advises companies and I spent the last 25 years in agencies and companies telling stories about what they do from all the different angles; I founded the first hybrid production agency in Denmark when video arrived to the web (Does anyone remember Flash?), then via merger became Chief Digital Officer of a full service agency catering the digital strategies and productions, and finally flew to Silicon Valley where I spent a decade building out Netlify, a web development platform that today gets more than 5000 new developers per day and runs 50 million stores, sites and apps.

I've gone from having to fight for Digital having a seat at even the smallest table, to witnessing it becoming close to everything, I've had to rethink anchored positions again and again and found that not only is change the only constant, but that it seems to be ever accelerating as well.

Before digital the channels we used to promote our brands - TV, print, outdoor, events and direct mail - were king for many decades. And while the rules for some brands might have slightly altered along the way (has anyone seen clips from the NBC Camel Newshour where the presenter was literally flanked on both sides by stacks of cartons of Camel cigarettes?) generally speaking the channels were fairly set, and your marketing investments could be made with the ability to have an at least fairly good view around the corner.

Then digital arrived. First off it essentially meant throwing a business card online. A static website linking to the company's real-world business. It was barely a registrable line item in the budget.

Then we started adding things like AdWords and banner ads, SEO became a discipline and with Web 2.0 we saw the arrival of interactive websites. Digital was becoming harder to ignore and understanding the basic premises was becoming more essential for marketers.

The end of the 0's was also the first time we had to shed our skin. The game was changing too much and too fast for us to keep what we had. The static websites and banner ads were no longer enough, and with the emergence of Mobile and social media we started having a ton of new channels available. Content marketing, where we in place of carpet bombing our advertising messages, turned to actually demonstrating and qualifying our products, was quickly becoming more essential as new homes like YouTube really started to take off as well. Not to mention that the need for added scale and flexibility was starting a joint transition to the cloud.

This translated into the first time of being sold the 'end all' digital transformation! Abandon your old Monoliths in favor of these shiny new ones! However, it was a shallow promise, and indeed today we've learned that digital transformation is probably the only thing you can count on as being forever ongoing.

And it wasn't just marketing that was having to adapt rapidly. Companies today have long been digital businesses. In fact, I'd argue that there is a strong case to be made that most business problems today have digital solutions. But this is when that started to happen. Our business backbone was strongly digitized and ERPs, CRMs, and countless other enterprise software were becoming dominant line times in all enterprise budgets.

Together with the explosion of channels and opportunities, engaging with our audience online was becoming essential. This led to an increasing need for faster iteration. If you could only release new updates every few months, you were almost losing by default.

This was also why I co-founded Netlify. A new architecture allowed for changes being made easily and many times a day instead of once per quarter. It meant introducing agility into your digital systems by enabling a change of your services to composable and headless throughout. It gives you the desperately needed flexibility to separate your touchpoints like apps, stores and sites, from your backend data which in turns means being able to not have to change entire (and very costly) systems all at once, but gradually along the way. Something that is more relevant today than ever, as the no. of digital services available has exploded.

We now have 10 times more digital services at our disposal than 10 years ago, and back then we had 10 times more than 10 years before that! And with AI suddenly being everywhere we're once again seeing a complete rollover.

In the beginning the focus has been on generative AI. To a vast extent I believe this to be a direct continuation of the same story that has been unfolding since the beginning of software, which is abstraction. We're automating away something that used to be manual.

A few decades ago, most developers knew machine code. Today that has been abstracted away for almost all developers. And the same is happening for processes like fact checking, email writing, code writing, you name it.

However the buck - of course - won't stop there. Because just as the abstraction away from machine code didn't lead to less developers (on the contrary!) AI will lead to more. Much much more. More channels, more content, more everything.

For marketers, one thing that will be especially impacted is personalization. When it comes to web properties 100% of the Fortune 500 have invested in personalization software, though 40% don't utilize it at all. It's a typical scenario where large vendors (no one named, no one shamed) were selling functionality where the value prop was self-evident, but in practicality the software couldn't deliver.

However everything indicates that this is about to change fundamentally. I think our digital touch points will go from being interactive to being outright conversational. AI will enable us to present 100,000 unique experiences to 100,000 visitors instead of just one. This will require a whole new way of thinking about our brand, on how singular the perception of it should be, on what needs to stay the same for all (eg Hermès needs to remain exclusive regardless of what customer they speak to and where), and what can change to create maximum resonance with the specific individual.

I also see a big challenge in the aggregators becoming ever more powerful. There are already examples of it. To check the weather Google no longer leads you to weather.com, but shows it to you right there in the search results, and we're seeing an onslaught of social commerce, where the entire customer transaction takes place on platforms brands do not control. In a world where we'll see less of our customers using our owned media, we will have to find new ways of creating relationships and installing what our brand promise is, to differentiate on anything else than price. Remember, it all looks the same on Amazon...

The onslaught of AI is also example of something where we can basically count its age in months yet we're already at a place that if you as a company are not leaning in you are finding yourself at a competitive disadvantage.

However you cannot tell me that at this early stage where AI is redefining its reach and limits on almost a weekly basis that anyone is able to pick out the clear winning approach. And that presents a dilemma. How do we pick our winners?

Some readers might say that at the end of the day it's all just more of the same. Whether it's using one CMS or another, this social media platform or that, it's, as we'd say in Danish, 'old wine in new bottles'. While I won't argue that there is some truth to this, I'm also seeing real change. When it comes to digital solutions, back when I was in agencies recommending these, every conversation I had with clients - whether they were CMOs or CTOs - would at the end of the day always have the same theme. They were attempts to find out what was the right stack or solution to choose. However, when I have these conversations today, gone is the notion that there exists *a* right solution. It's been replaced by a general acceptance that digital transformation is, and forever will be ongoing, that their digital landscape at scale will always have both new and legacy elements, and there is now a much larger focus on choosing tools and creating workflows that allow the organisation to be agile; To be able with as little friction as possible to for example deploy that business intelligence tool into current stacks and to discontinue it again. The combined abundance but also importance of tools, solutions and channels creates overhead and is the reason that Time to Market has risen to become the number one concern for most CMOs today.

For better or worse there is no respite in sight with entirely new channels like AR (Augmented Reality) being just around the corner! And it's not just how we go to market that is changing rapidly these years. The same goes for many other disciplines. In the last few years Covid forced us to introduce remote work at scale which came with an army of new collaborative software but also brand new workflows within our teams. Distributed cultures have their own challenges that we must adapt to and overcome as well and we've now by long ago entered an age where everyone is in marketing (almost everyone has a social media profile).

To say that our brands—both externally and as an employer— have to be elastic would be an understatement of the year.

By now you've probably guessed where I'm going with all this. Change is happening at an ever-increasing rate, and the only way to take advantage while staying afloat is to adopt an agile approach throughout your organisation and your workflows. And how to both adapt to that mindset and implement this in practice is exactly what this book is all about. Enjoy reading!

Chris Bach
San Francisco
12/26/2024

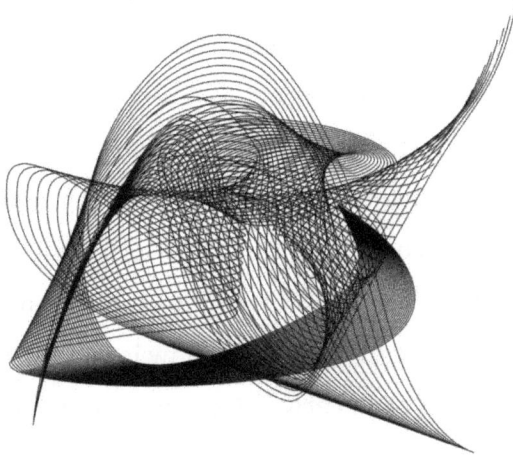

Introduction to the Revised Edition

"A brand is no longer what we tell the consumer it is—it is what consumers tell each other it is."
—*Scott Cook*

Seven years ago, as of the publishing of this book, I set out to capture my thinking based on over a decade of working with some of the world's top brands with my digital experience agency while also "coming up" in the marketing and technology industries in a time when the Internet became mainstream, social media was both invented and ultimately became a force in the marketing world, and where, more recently, smartphones replaced desktop computers as consumers' go-to device, customer data privacy became a big issue warranting tough regulations, and the term "customer experience" was just starting to come into its own in the corporate world.

At this intersection, along with my own personal journey in adopting more agile methods of managing, working, and thinking, I wrote the first edition of The Agile Brand, which came out in July 2018. I am still thrilled to this day at the reception it received, and based on that, I was inspired to start my podcast, now in its sixth year with millions of downloads and nearly 600 episodes as of the publishing of this latest edition.

But obviously, a lot has changed since I wrote the first edition of this book back in 2018. So I decided to revisit it entirely, in some cases completely removing things that felt a little obvious or outdated and adding in completely new sections that feel more aligned with our times. This book is the result of that effort.

For instance, in 2018, the concept of customer experience as a discipline and core focus of business resources was a more novel one. Nowadays, this is commonplace, so I've incorporated these ideas not as a nascent concept but as an understood paradigm. There are many great resources that support this with statistics, survey results, and company financials, and I encourage you to seek those out if you remain skeptical about the value of a customer experience focus for an enterprise.

As you can imagine, many other things have changed since my pre-COVID, pre-AI peak hype book was originally published. So, I've made sure to make the text that follows a reflection of our current times.

Who this book is for

This book is for marketers and those in customer-focused roles. While I speak primarily to the marketers out there, other roles, such as designers, customer experience professionals, and others, can likely gain value from the ideas in this book. Additionally, as I talk about the need for greater alignment across the business towards things like a shared definition of business value, there is something here for any business leader, regardless of where they sit within the organization.

What this book will cover

I have separated this book into three parts and followed the same general structure as its original version. Here is what we will explore together:

Part 1: The Fundamentals

In this section, we lay the groundwork by exploring the essential concepts and principles that underpin the Agile Brand. We begin with a comprehensive overview of what it means to be an Agile Brand and why agility is crucial in today's dynamic market landscape. You'll learn about the key drivers of brand agility, including the importance of customer-centricity, continuous improvement, and strategic flexibility.

This foundational understanding will set the stage for deeper insights into the specific characteristics and practices defining successful Agile Brands.

Part 2: Defining the Agile Brand

Here, we delve into the core values and principles of the Agile Brand framework. We introduce the seven principles of an Agile Brand: Agility by Design, Continuously Improving, Operationalizing Adaptivity, Guided by Values, Building Relationships, Focusing on the Conversation, and Always Learning and Growing. Each principle is examined in detail, providing insights into how they contribute to brand agility and success. You'll also learn how to apply these principles in your organization to foster a culture of agility and resilience.

Part 3: Building the Agile Brand

In the final section, we move from theory to practice, focusing on the strategies and tools needed to build and sustain an Agile Brand. You'll discover practical steps for implementing a feedback loop, creating a culture of experimentation, and aligning your organization with the core values of substance, focus, and relevance. We'll also cover how to measure and evaluate your brand's agility, ensuring continuous improvement and long-term success. Real-world examples and case studies illustrate how leading companies have successfully adopted

Agile Brand principles, providing you with actionable insights to apply in your own organization.

What this book is not

This is not intended to be a book teaching the principles of branding. There are many really great books on this topic, and I am not a branding expert per se, although I've worked on award-winning branding initiatives and certainly worked with some of the world's leading brands. That said, this book is about how to create the essence of a brand internally and with our customers that those branding experts can then build the visual and messaging "brand" upon.

We'll take a look at the evolution of branding from the lens of a customer relationship with a brand. Still, it is not my intention for this book to serve as a guide on how to write a vision or mission statement, craft a good tagline, design a memorable logo, or anything of that nature.

Additional resources

My podcast, *The Agile Brand with Greg Kihlström,* is a great companion to this book. On the show, I talk with business leaders and platform experts who talk about how to practically apply many of the principles discussed in this book. I highly recommend you follow it on your podcast platform of choice.

Additionally, you can find many more resources on The Agile Brand Guide website at https://www.agilebrandguide.com

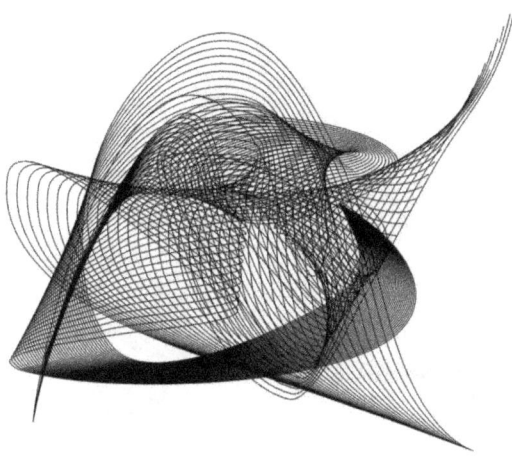

Part 1: The Fundamentals

"Branding is the art of differentiation."
—David Brier

As I already said in the introduction, my purpose with this book is not to write another primer on branding. There are many great books on that topic. That said, to proceed to the topics I want to cover later in this book, we do need to set the stage a bit, so to speak.

So, in this section, we will explore the four tenets of an Agile Brand, the values that comprise one, and look at what we mean by the terms *agile* and *agility*.

Then, we will walk through how I see the evolution of brands and branding, which underscores why creating an Agile Brand is so important.

So, let's briefly cover the fundamentals here.

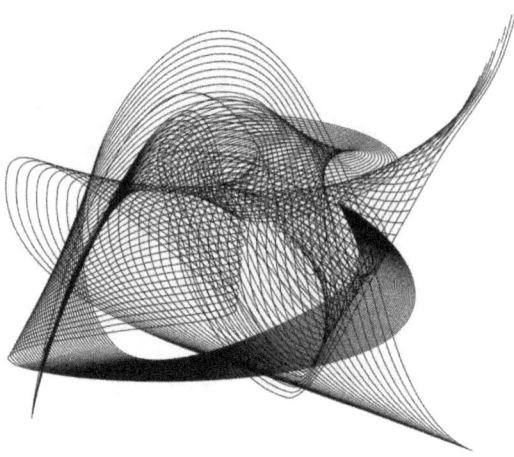

Chapter 1 |
What Is an Agile Brand?

TL;DR

A brand is more than just a logo; it is a living, evolving relationship with consumers, built on a unique vocabulary and genuine interactions. The core components of a strong brand are substance, focus, and relevance. These elements ensure that a brand stands for something meaningful, targets the right audience, and stays connected to changing consumer needs and preferences.

As you likely know, there are many components of a brand. Things like a vision, mission, logo, tagline, and other elements are surely covered in other books specifically about branding. However, since this book is focused on a specific aspect of branding, we will leave those items for another book. We are here to talk about an *Agile* Brand, so what we will talk about in this chapter builds on the basics of what a brand consists of and focuses our attention on what it means to create an Agile Brand.

The Four Tenets of an Agile Brand

So, to explore what an Agile Brand is, we need to have some basic rules we can apply. This brings us to what I call the four tenets of a brand. Let's explore.

4 Tenets of an Agile Brand

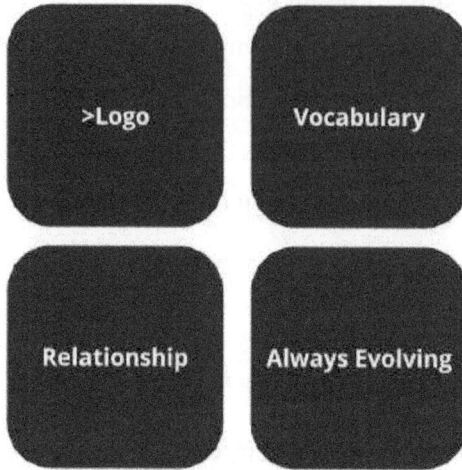

Figure 1.1, The four tenets of an Agile Brand

Tenet 1: Your Brand Is More Than a Logo

In the modern marketplace, understanding the multifaceted nature of branding is crucial. A logo, while a significant symbol, is merely one component of a brand's identity. The true essence of a brand extends far beyond a visual mark. It encompasses an intricate tapestry woven from various elements, including digital presence, social media personality, and content strategy.

For instance, Apple's logo is iconic, but what truly defines Apple is the seamless experience it provides across its products and services. From the intuitive design of its devices to the user-friendly ecosystem that connects them, Apple's brand identity is a holistic experience that goes far beyond its bitten apple logo.

Tenet 2: Your Brand Is a Vocabulary

A brand is a vocabulary that communicates the essence of a company to its audience. This vocabulary comprises the words, images, and messages that convey the brand's values, mission, and unique selling propositions. It's the language that customers use to describe and relate to the brand.

Consider Google. The term "Google" has become synonymous with searching the internet. This linguistic shift illustrates the power of a brand vocabulary that not only defines a company but also integrates into the everyday language of its users. Similarly, brands like Kleenex and Coke have become generic terms for tissues and soft drinks, respectively, showcasing the pervasive influence of a strong brand vocabulary.

Tenet 3: Your Brand Is a Relationship

Brands today are living, breathing entities capable of forming deep, interactive relationships with consumers. These relationships go beyond transactions; they are built on trust, engagement, and mutual value. Social media and digital platforms have transformed how brands interact with customers, making these relationships more dynamic and reciprocal.

Nike's relationship with its customers exemplifies this tenet. Through initiatives like Nike+, the brand has created a community of runners who connect, compete, and share their experiences. This relationship extends beyond purchasing running shoes; it encompasses a lifestyle and a sense of belonging to a global community.

Tenet 4: Your Brand Is Evolving – Whether You Intend It to or Not

We live in an age where brands are constantly evolving, and I can't imagine this will stop anytime soon—or any time at all. This evolution is driven by changing consumer preferences, technological advancements, and societal shifts. Whether a brand actively manages this evolution or not, it is continually shaped by external and internal forces.

Netflix provides a compelling example of brand evolution. Originally a DVD rental service, Netflix transformed itself into a streaming giant and content creator. This evolution was not just about adapting to new technology but also about redefining the brand to align with the changing ways people consume media.

The Three Values of an Agile Brand

In the journey toward becoming an Agile Brand, three core values serve as guiding pillars: Substance, Focus, and Relevance (Figure 1.2). These values are essential in shaping a brand that is not only agile and adaptive but also deeply resonant with its customers and stakeholders.

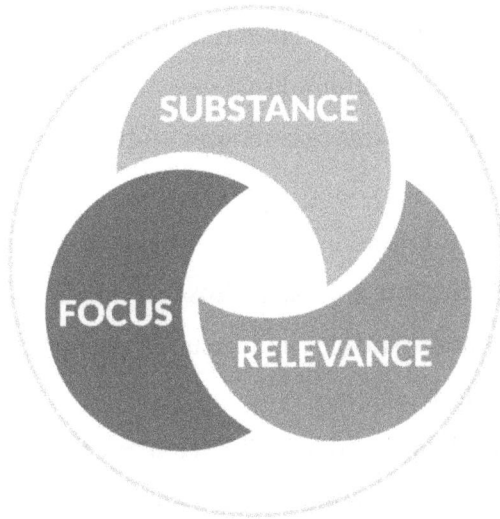

Figure 1.1, The three values of a strong brand

Substance

Substance is the foundation of a strong brand. It encompasses the core values, mission, and genuine experiences a brand offers its audience. A brand with substance stands for something meaningful and consistently delivers on its promises.

Substance refers to the depth and authenticity of a brand. It encompasses the brand's core values, mission, and genuine commitment to delivering quality and meaningful experiences. A brand with substance is built on a foundation of integrity and purpose, ensuring that every action, message, and product aligns with its fundamental principles.

Patagonia is a prime example of a brand with substance. Founded on the principles of environmental sustainability and ethical business practices, Patagonia's commitment to these values is evident in every aspect of its operations. From its product designs to corporate activism, Patagonia's substance resonates with consumers who share its values.

To build substance:

- **Articulate a Clear Philosophy:** Define what your brand stands for and ensure it permeates all aspects of your business.
- **Deliver Authentic Experiences:** Create meaningful interactions reinforcing your brand's values and promise.
- **Maintain Transparency:** Be open about your processes, successes, and challenges to build trust and authenticity.

In a world where consumers are increasingly seeking authenticity, substance sets a brand apart by providing real value and building trust. It ensures that the brand's promises are kept and that its contributions are meaningful and impactful.

Focus

A successful brand must have a clear focus. This focus defines its target audience, unique selling propositions, and strategic direction. Brands that try to appeal to everyone often fail to resonate with anyone.

Focus involves the brand's ability to concentrate its efforts and resources on what truly matters. It means clearly understanding the brand's core strengths and strategic priorities and directing all activities towards achieving these goals without being distracted by extraneous opportunities.

Apple's focus on simplicity and premium quality has been a cornerstone of its success. By targeting a specific audience that values design, innovation, and seamless user experience, Apple has maintained its position as a leader in the tech industry.

To maintain focus:

- **Define Your Target Audience:** Understand who your primary customers are and tailor your brand message to their needs and preferences.
- **Highlight Your Unique Selling Propositions:** Clearly communicate what sets your brand apart from competitors.
- **Stay Consistent:** Ensure all brand communications and actions align with your core focus.

Focus enables a brand to be efficient and effective, ensuring that it delivers on its promises with precision and excellence. It allows for strategic clarity and consistent execution, which are crucial for maintaining a strong brand identity and achieving long-term success.

Relevance

Relevance is about staying connected to the needs and desires of your audience. A relevant brand adapts to cultural shifts, technological advancements, and changing consumer behaviors without losing sight of its core identity.

Relevance is the brand's ability to stay pertinent and significant in the eyes of its audience. It involves continuously evolving and adapting to meet the changing needs, preferences, and expectations of customers. A relevant brand stays connected to its market and remains attuned to societal trends and technological advancements.

Lego is a brand that has remained relevant across generations. By continually innovating and expanding its product offerings to include digital games, movies, and themed sets, Lego has kept its brand fresh and appealing to both children and adults.

To ensure relevance:

- **Monitor Trends:** Stay aware of changes in consumer preferences and industry trends.
- **Adapt Strategically:** Implement changes that enhance your brand's appeal while staying true to its core values.

- **Engage with Your Audience:** Regularly interact with your customers to understand their evolving needs and expectations.

In an ever-changing marketplace, relevance ensures that a brand remains top-of-mind and competitive. It fosters ongoing engagement and loyalty by ensuring the brand's offerings and messages resonate with current and future customers.

Conclusion

A brand is much more than a name or logo; it is a complex, evolving relationship built on a foundation of substance, focus, and relevance. By understanding and embracing these tenets and values, brands can create enduring connections with their audiences and remain impactful in an ever-changing marketplace.

Next Best Actions

1. **Define Your Brand's Core Values and Mission:** Reflect on what your brand stands for and articulate its core values and mission. Ask yourself: What do we believe in? How do our products and services reflect these beliefs?

2. **Identify and Understand Your Target Audience:** Conduct research to pinpoint your primary customers. Develop detailed personas and ask: Who are our customers? What are their needs, preferences, and pain points?

3. **Evaluate and Enhance Your Brand's Relevance:**
   Regularly assess your brand's relevance in the market.
   Monitor trends and engage with your audience through
   surveys or social media. Ask: How can we adapt to stay
   connected to our customers' evolving needs? What changes
   can we implement to remain fresh and appealing?

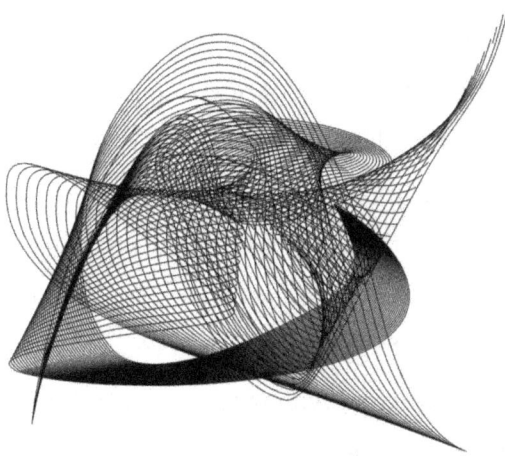

Chapter 2 |
Agile and Agility

TL;DR

Understanding the difference between formalized
Agile principles and the broader concept of
agility is crucial for modern businesses. While
Agile provides a structured framework for
efficiency and collaboration, true agility is about
being nimble and adaptive to change. Brands
must balance adherence to Agile principles with
the flexibility to respond proactively to evolving
market conditions and consumer needs.

In the fast-paced world of modern business, the terms "Agile" and "agility" are often used interchangeably. However, while they are related, they represent distinct concepts. Agile refers to a specific set of principles and methodologies designed to improve efficiency and collaboration, primarily in software development but increasingly in other fields. Agility, on the other hand, is the broader ability of a business to remain flexible, responsive, and adaptive to change. This chapter explores the difference between formalized Agile principles and the concept of agility, providing insights into how brands can leverage both to stay competitive.

Let's Start with a Recap of What Agile Is

To continue our discussion of what we mean by an "Agile brand," we need to explain exactly what we mean by the term Agile. Merriam-Webster defines it as[1]:

> 1: marked by ready ability to move with quick easy grace: an *Agile* dancer
>
> 2: having a quick resourceful and adaptable character: an *Agile* mind

While these are not too far off, when I refer to Agile in this book, I am referring to the Agile principles that a group of software engineers agreed upon a few decades ago and which set a number of things in motion, including the ideas in the book you are reading right now!

For reference, I've included the 12 Agile principles in the appendix of this book with a little more commentary if you want to refer to them later. At a high level, they are the following:

- Principle 1 is that the highest priority is satisfying the customer.
- Principle 2 is that Agile processes harness change for the brand's competitive advantage.
- Principle 3 is to deliver working output frequently, with a preference for a shorter timescale.
- Principle 4 states that business people and developers—or, in this case, marketers—must work together daily throughout the project.
- Principle 5 is focused more on the teams doing the work. It is to build projects around motivated individuals.
- Principle 6 states that direct conversation is the most efficient and effective method of conveying information.
- Principle 7 states that working output is the primary measure of progress.
- Principle 8 states that Agile processes promote sustainable development, work, and the most potential for ongoing improvement.
- Principle 9 states that continuous attention to technical excellence and good design enhances agility.
- Principle 10 states that simplicity—the art of maximizing the amount of work *not* done—is essential.

- Principle 11 states that the best architectures, requirements, and designs emerge from self-organizing teams.
- Principle 12 states that the team reflects on how to become more effective at regular intervals.

To understand why it was important to define these 12 principles, it helps to consider what the world was like *before* Agile.

But first, there was waterfall...

The waterfall methodology (Figure 2.2.1) was first used in the early 1970s, after a decade of SDLC being the dominant methodology used in the software development community.

REQUIREMENTS

SYSTEM DESIGN

IMPLEMENTATION

TESTING

DEPLOYMENTS

MAINTENANCE

WATERFALL METHOD

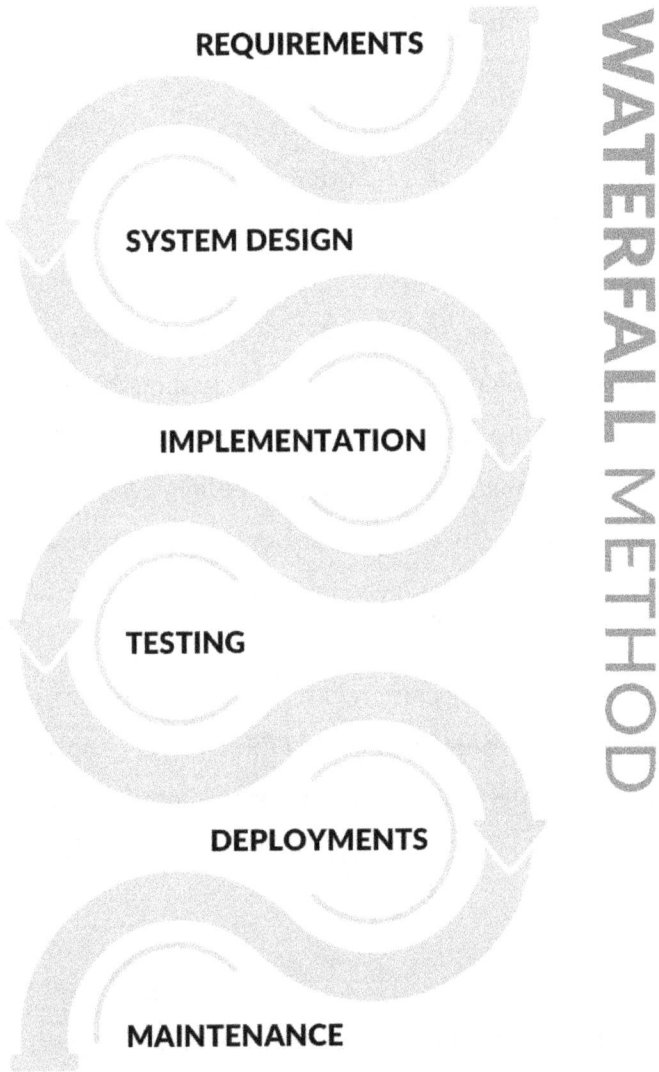

Figure 2.1, Waterfall Method

Although there are variations among the exact steps and names various people or companies use in the waterfall process, the general process is consistent. The central idea is that waterfall is a very linear process.

Changes during the latter phases of a waterfall project will often require major rework, repetition of testing, and other downstream activities.

In other words, if you get all the way to the "Implementation" step (the 3rd step in the chart above) and realize you missed something or that you need to change the way something is currently created, it will require you to start from the beginning and re-integrate these new requirements, as well as a new design, and then to re-implement the code. With waterfall, it can be very costly and time-prohibitive to make changes, even if they massively improve the end product.

In addition, early computers were large, expensive, and almost impossible to obtain for long amounts of time. With a short amount of time to run your programs, you had to spend a lot of time ensuring you got things right.

Introduction of Agile

After decades, the waterfall method, the availability of fast computing, the rise of the Internet, and the ubiquity of personal computers allowed more rapid development of software and, eventually, web applications.

While several methodologies popped up earlier in the 1990s, including Rapid Application Development[2] (RAD), Agile development was the most effective at "winning over" software developers.

In 2001, the Agile Manifesto[3] was born:

"We are uncovering better ways of developing software by doing it and helping others do it. Through this work we have come to value: Individuals and interactions over processes and tools

Working software over comprehensive documentation Customer collaboration over contract negotiation Responding to change over following a plan

That is, while there is value in the items on the right, we value the items on the left more."

Agile centered around the belief in collaboration, iteration, and a more "social" (not to be read as "social media") approach to creating great products.

Scrum, among others

While Agile itself is not a methodology, the Agile principles outlined in the Agile manifesto were then used to create several methods of doing work in an Agile manner. One of the more popular methods is called Scrum, and it became so popular that "Scrum" and "Agile" are often interchangeable for some.

For the purposes of this discussion, those terms might be used interchangeably in the paragraphs that follow, though keep in mind that there are other methods that use Agile principles such as Kanban, SaFE, and more.

How Scrum works

An Agile Scrum software development "Sprint" (Figure 2.2.2) consists of several iterations over a period of days or weeks that is driven by product requirements.

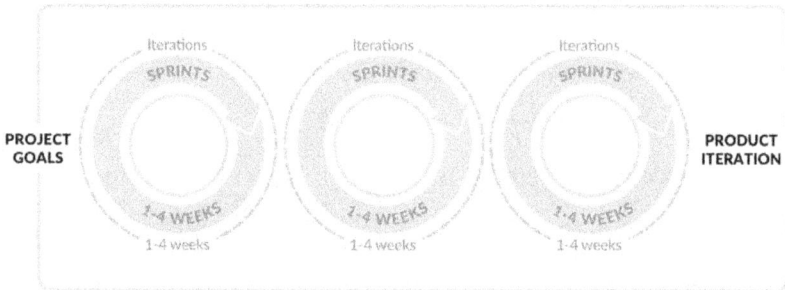

Figure 2.2, Sprints as performed using Scrum methods

Unlike the linear steps of the Waterfall methodology, the Scrum methodology uses an iterative process to create products such as software. Through interactions that focus on specific goals and product features, called sprints, the product incrementally improves as it works against a product backlog that contains the features the end product must contain.

A Scrum project, using Agile principles, completes several such sprints throughout the project lifecycle, as opposed to a single development period in the waterfall method. Thus, a Scrum project consists of multiple sprints that end in a new version of the product.

Agile's rise and subsequent contributions to the world of software and the Web at large are wide-reaching and have changed how we create and market products and services.

Principles vs. Dogma

Agile principles, outlined in the Agile Manifesto, emphasize customer satisfaction, collaboration, and responsiveness to change. These principles guide teams in delivering high-quality work iteratively and incrementally. However, it's essential to distinguish between adhering to these principles and becoming dogmatic about specific Agile methodologies.

Being too rigid in following Agile methodologies, such as Scrum or Kanban, can hinder a team's ability to respond effectively to unique challenges and opportunities. For instance, while daily stand-ups and sprint planning are valuable, they should not become burdensome rituals that stifle creativity or slow down progress. The true essence of Agile lies in its principles, not in the dogmatic application of its practices.

Principles vs. Dogma in Action

Principle	Dogma
Customer Collaboration Over Contract Negotiation	Insisting on rigid adherence to contracts, even when it's clear the customer's needs have changed.

GREG KIHLSTRÖM | 40

Responding to Change Over Following a Plan	Refusing to deviate from a predefined plan, even when new information suggests a better approach.
Working Software Over Comprehensive Documentation	Producing minimal documentation that fails to support the software effectively simply to claim adherence to Agile.
Individuals and Interactions Over Processes and Tools	Mandating the use of specific tools or processes, even when they do not suit the team's workflow.
Simplicity – The Art of Maximizing the Amount of Work Not Done	Cutting corners in the name of simplicity, compromising quality and long-term maintainability.

Figure 2.3, Principles versus Dogma in Action

Being Agile vs. Being Reactive

There's a significant difference between being agile and being reactive. An agile organization proactively anticipates changes and adapts its strategies and processes to meet new demands. In contrast, a reactive organization only responds to changes after they occur, often scrambling to catch up.

For example, an Agile Brand might continuously monitor market trends and customer feedback to innovate its products and services, staying ahead of competitors. A reactive brand, however, may only implement changes after losing market share or facing customer dissatisfaction. True agility involves foresight and strategic planning, ensuring a brand can pivot smoothly and efficiently in response to emerging trends.

Being Agile vs. Being Reactive in Action

Being Agile	Being Reactive
Proactively monitoring market trends and customer feedback to anticipate changes and innovate products.	Waiting until market share declines or customer complaints escalate before making changes.
Implementing iterative processes that allow for regular assessment and adjustment of strategies.	Only making adjustments after encountering significant issues or failures in existing strategies.
Encouraging cross-functional collaboration to foster innovation and address potential challenges early.	Operating in silos and only involving other departments when problems become too large to handle alone.
Empowering employees to make decisions and adapt processes as needed to stay aligned with business goals.	Requiring all decisions to go through multiple levels of approval, causing delays and missed opportunities.

Investing in ongoing training and development to ensure the team can quickly adapt to new technologies and methodologies.	Providing training only when a new system or technology is already causing operational issues.

Figure 2.4, Being Agile Versus Being Reactive in Action

Embracing Agility and What It Means for Brands

Embracing agility means fostering a culture of continuous improvement, flexibility, and responsiveness. For brands, this translates to several key practices:

1. **Customer-Centric Approach:** Continuously gathering and acting on customer feedback to refine products and services.

2. **Cross-Functional Collaboration:** Encouraging teams across different functions to work together seamlessly, breaking down silos and enhancing innovation.

3. **Iterative Processes:** Implementing iterative cycles of planning, executing, and reviewing to enable quick adjustments and improvements.

4. **Empowerment and Trust:** Empowering employees to make decisions and trust their judgment to respond swiftly to changes.

Netflix exemplifies agility by consistently adapting its content strategy based on viewer preferences and data analytics. Its ability to quickly pivot from a DVD rental service to a streaming giant and then to a major content creator showcases its commitment to staying agile in a rapidly changing industry.

Things to Avoid

While striving for agility, brands should be mindful of potential pitfalls:

1. **Over-Reliance on Tools:** Tools and software can facilitate Agile practices but should not become the focus. The emphasis should remain on people and interactions.
2. **Neglecting Long-Term Vision:** In the pursuit of short-term agility, brands should not lose sight of their long-term goals and vision.
3. **Resistance to Change:** Agility requires a willingness to change. Brands resistant to change, either culturally or operationally, will struggle to remain agile.

Conclusion

Balancing formalized Agile principles with the broader concept of agility is key to a brand's success in today's dynamic market. By understanding the difference between being agile and merely reactive and by fostering a culture that embraces flexibility and continuous improvement, brands can remain competitive and resilient.

Next Best Actions

1. **Evaluate Your Agile Practices:** Review your current Agile methodologies and assess whether they enhance your team's flexibility or are becoming rigid routines. Ask: Are our Agile practices helping us adapt quickly and efficiently?

2. **Foster a Culture of Continuous Improvement:** Encourage feedback and innovation at all levels of the organization. Implement regular reviews and retrospectives to identify areas for improvement. Ask: How can we create an environment that supports ongoing learning and adaptation?

3. **Enhance Cross-Functional Collaboration:** Break down silos between departments and promote teamwork across functions. Encourage open communication and collaboration. Ask: How can we better integrate our teams to respond more effectively to changes?

By focusing on these actions, brands can develop the agility needed to thrive in an ever-changing marketplace.

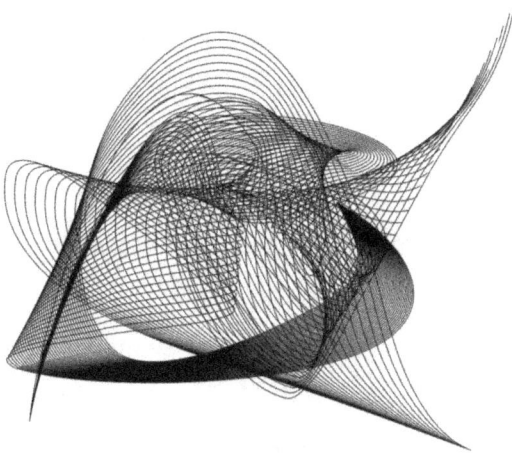

Chapter 3 |
The Evolution of Brands

TL;DR

There are five stages in the evolution of brands, which takes us from the simplest incarnation of a mark that represents a company and its products to a collaborative relationship with a brand, facilitating customers to have amazing experiences.

1. Brand as object
2. Brand as idea
3. Brand as experience
4. Brand as relationship
5. Brand as facilitator

Being a great brand marketer means fully understanding how consumers interact with and experience brands. It also means you know and appreciate how the discipline has evolved over the years, as well as how consumers interact with companies, products, and causes. While some fundamentals never change, it's important to understand how brands have evolved in the eyes of brands and their audiences over the years. This also gives us the context to understand where brands are headed in the future and their evolving relationship with their customers.

The evolution of brands encompasses five eras, each of which we'll review in more detail:

1. Brand as object
2. Brand as idea
3. Brand as experience
4. Brand as relationship
5. Brand as facilitator

As you work tirelessly to create an engaging experience between your audiences and your brand, keep this evolution in mind. You can also look at these as four dimensions within which your brand needs to exist. Remember that none of these dimensions have become irrelevant, even as brands' relationship with consumers has evolved. Instead, it's an additive process. Over time, brands have become more complex, with each new evolution adding a new layer of meaning and understanding. Now, let's get started.

Brand as object

BRAND AS OBJECT

AWARENESS　　　　**ACTION**

Figure 4.1, Brand as object: Name, company, and/or product recognition translates to sales

Our use of logos to represent companies, organizations, or individuals is based on a long history that originates with the very beginning of written communication. The moment we used a drawing, an image, or a symbol to represent *something else*, we'd been essentially branding things. From cave paintings to hieroglyphics to the first logo representing a company, we've been using graphic representations to give meaning to ideas.

Dan Redding puts it this way[4]:

"Signs can take the form of words, images, flavors, or even odors: things that have no intrinsic meaning until we invest it in them. We perceive, understand, and negotiate the world around us by investing meaning in all manner of signs and symbols. In the West, an image of a snake signifies evil. But without our Western cultural and mythological associations

(many of which are rooted in the Bible), a serpent is just a serpent."

Figure 4.2, The Stella Artois logo, first used in 1366, is the earliest known logo.

In the early days of marketing and advertising, it was enough to simply have a name or product recognition in order to generate sales. Mass advertising created mass sales.

Brand as idea

With increasing competition from mass production, mass advertising, and mass media in general, the need for brands to be more than an object came along. At this point, it wasn't enough to simply have a recognizable logo and a product that was available in stores.

BRAND AS IDEA

AWARENESS　　　**PERCEPTION**　　　**ACTION**

Figure 4.3, Brand as Idea: Perception of the products or company and the feelings and ideas they evoke translate to sales

Brands now needed to compete for mindshare, or as Will Burns says in his article on Forbes.com, "Branding is what happens to people after they've spent time interacting with your company[5]."

As consumers and marketers have grown more sophisticated, it is not enough to simply be known. Companies and products must stake out a claim on an area of the popular imagination and exist as an "idea."

For instance, seeing the "golden arches" of McDonald's plants an idea in your head. Maybe it's a Big Mac, fries, or a large Coca-Cola. Whatever it is (and whatever your feelings about its products), that yellow "M" brings with it more than just an opinion of yellow on red typography.

The same applies to Starbucks, which was so successful with its previous branding recognition efforts that it removed the word "Starbucks" from its logo and suffered no discernable setbacks.

Brand as experience

With increased competition and consumer preference for more tailored products and services, brands were forced to differentiate themselves beyond occupying an idea. They needed to insert themselves into key life moments and become part of our experience.

BRAND AS EXPERIENCE

| 1 | 2 | 3 | 4 |
| AWARENESS | PERCEPTION | ENGAGEMENT | ACTION |

Figure 4.4, Brand as Experience: Customers' interactions with brands build deeper connections, which translate to sales

Going back to an earlier brand example, Apple's brand experience extends all the way from the initial sale (either in one of its branded stores or its online presence), through the packaging you open to first use your product, through the ease of setup, through usage of the product every day. And if you have problems, you can go to the Apple Store to ask questions.

In more recent times, brands have used experience to cut through the clutter of marketing and advertising, creating deeper engagement with customers.

Brand as relationship

As more and more brands have adopted the experience approach, it has become clearer that a one-off moment in time is not enough to cement brand loyalty. This takes us to the current stage in the evolution of brands: brand as relationship.

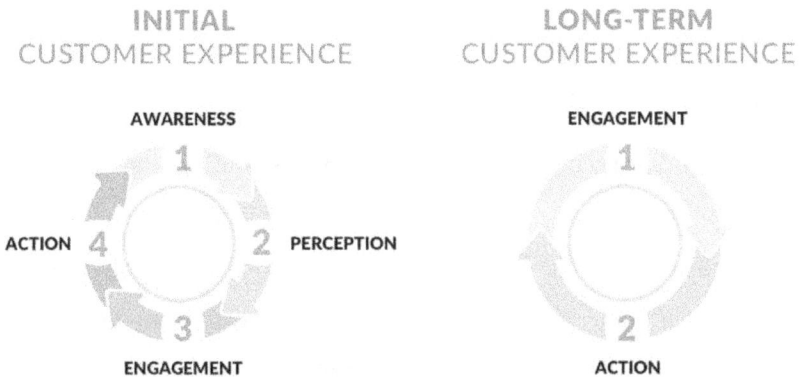

INITIAL
CUSTOMER EXPERIENCE

LONG-TERM
CUSTOMER EXPERIENCE

AWARENESS
1

ACTION 4 2 PERCEPTION

3
ENGAGEMENT

ENGAGEMENT
1

2
ACTION

Figure 4.5, Brand as Relationship

Mark Bonchek and Cara France have a great take on this in their article in Harvard Business Review[6]:

> To get started, think about the relationship people
> have with your brand today. Frame your answer as

social roles. For example, if you are a healthcare
provider, you probably have a brand relationship
based on doctor/patient. Now think about other kinds
of relationships outside your industry. For example, in
health care there are aspects of teacher/student (to
educate), coach/athlete (to motivate), or
guide/traveler (to navigate). Be sure to consider roles
that are symmetrical, like friend/friend,
neighbor/neighbor or co-creator/co-creator.

Now that just about every brand is on social media and has some type of interactive presence, it's not enough to simply be available and able to be searched and found. It's also not enough to have a one-off experience with a customer.

To truly build long-term value with a customer, you need to consider the type of relationship you can realistically have with your audiences and build on that. This takes understanding the type of relationship your customers want to have with *you* and designing an experience around it. It also takes a holistic approach to your customer experience that goes beyond single channels, devices, or events. By thinking in terms of a long-term relationship instead of short-term wins, your brand takes on a new life, and a cohesive multi-channel strategy takes on even more importance than ever.

According to a recent study by Epsilon[7], 80% of consumers said they are more likely to buy from a company if their experience is personalized. Your customers may not always be ready to purchase, but keeping them engaged will make them more likely to buy from you when they are.

It is also important to understand that, while a customer might be loyal, they might not always be in the "buying" mood. So, smart marketers have learned how to nurture customers and keep them in a state of engagement, even while they may not always be ready to take action. This state of engagement also means they will be primed to refer your brand to their friends and colleagues when the time comes. Word of mouth occurs from engaged customers and makes a huge difference to brands that are able to achieve a loyal following.

Brand as Co-Creator

Earlier, we talked about the four stages of the evolution of the relationship between brands and consumers. Starting with a very abstract concept of a company and product or service represented by a mark or logo and evolving to a more mutually beneficial relationship, this has evolved over centuries.

FACILITATOR-MAKER

| 1 | 2 | 3 | 4 | 5 |

| OBJECT | IDEA | EXPERIENCE | RELATIONSHIP | FACILITATOR-MAKER |

Figure 3.5, The Co-Creator or Facilitator-Maker Brand Relationship

I believe we're now entering a fifth evolution that actually started fairly soon after the fourth. We are, after all, in a time of accelerating change. So, it makes sense that all things, including the brand-consumer relationship, would evolve at an accelerating pace. This power dynamic can be rather simply illustrated in the chart below:

POWER-VARIETY & ACCESS

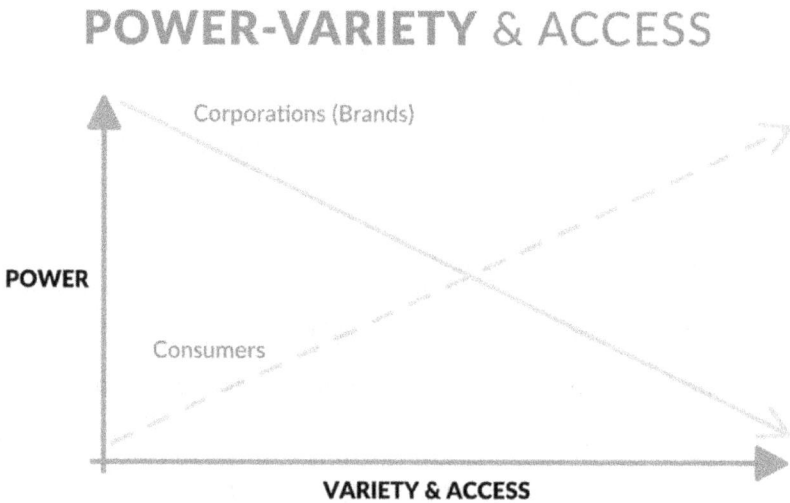

Figure 3.6, The Power Dynamic of Power-Variety and Access

As consumers have access to increasing varieties and options, and their agency to research and spend their money more wisely increases, the power that a few corporations or brands have over them decreases.

In this scenario, the brand, which was primarily thought of in the past as the "maker" of things, has now shifted to being a facilitator. The consumer is now the "maker," facilitated by brands. Consumers now have the power and the agency to choose what they want, when, where, and how they want it. Brands are there to provide services and facilitate the creation of consumers' experiences.

According to some Forrester Research as early as 2012, consumers aren't just willing co-creators; they are eager[8], and "61% of US online adults … are open to co-creating across a large range of industries," the study finds. However, a surprising "62% of all companies are not using social media to interact directly with their customers in order to influence product creation, design, or strategy."

So, not only is this not a new phenomenon or desire on the part of the consumer, but it is also something that continues to grow over time. We see this demonstrated in the number of ways that customization is offered on everything from computers to clothes to cars. We see it in the niche startup retailers that are able to capitalize on a narrow market. And we see it in our own desire as consumers to be a greater part of the creation process of new products and services.

Conclusion

You might also be wondering if all brands have made it to the more enlightened third, fourth, or even fifth stages. For better or worse, not all brands have, though the focus on customer experience of late means that most are embracing at least the third stage.

So, with this as background, it is now time for us to explore what exactly an Agile Brand is.

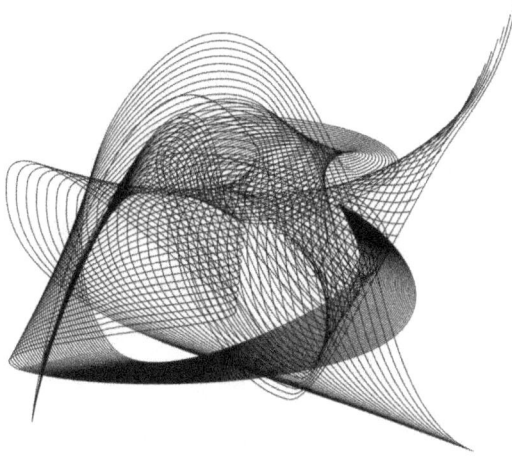

Part 2 |
The Principles of an Agile Brand

"A brand for a company is like a reputation for a person. You earn reputation by trying to do hard things well." —Jeff Bezos

Now that we have the fundamentals out of the way, we can explore in depth the ideas and principles behind what makes an Agile Brand unique, as well as the ideas behind it that make it compelling.

The Agile Brand Manifesto

As with the introduction of any concept, Agile branding needs a clear statement that unequivocally defines what it is and describes the philosophy that guides it. Much as Agile Manifesto did that for programming, I'd like to do that for Agile branding now.

I want to thank and give credit to the team that wrote the original Agile Manifesto for its inspiration. You can find all their names at AgileManifesto.org.

Through the evolution of brands from a simple visual indicator of ownership to their broad function today, we understand that a brand can be greater than the individual or individuals who created it, the teams who maintain it, and the products and services they represent.

Because of this, we have come to value the following:

- Long-term customers over short-term sales.
- Dialogue with customers over broadcasting one-way marketing messaging.
- Staying true to our values over doing whatever we can to generate profits.
- Continual improvement over maintaining the status quo.

While the constructs on the right remain important, we value those on the left more.

We also know that for a brand to be successful, it must open itself to consumers for feedback, ideas, and dialogue. No longer can brand decisions be made solely in a boardroom or by shareholders.

Consumers want and need to feel a connection with brands for them to be truly successful.

The 7 Principles of an Agile Brand

The rest of this section will explore what I have defined as the seven principles of an Agile Brand, which are the following:

1. Agility by design
2. Continuously improving
3. Operationalizing adaptivity
4. Guided by values
5. Building relationships
6. Focusing on the conversation
7. Always learning and growing

Let's continue our journey!

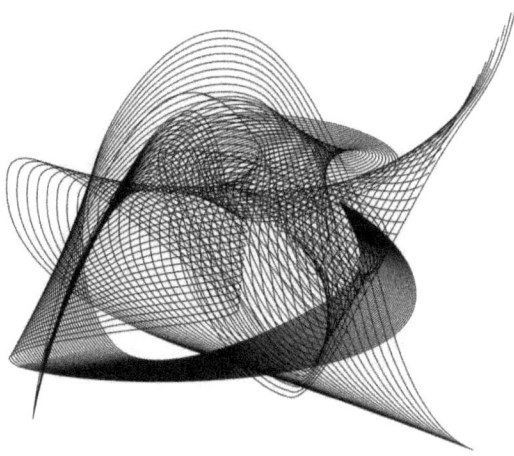

Chapter 4 |
Principle 1: Agility by Design

**Understand that change is the only constant;
build to change.**

Agility by design means creating a business structure and culture that inherently supports flexibility and rapid response to change. This principle is about embedding agility into the very fabric of the organization, ensuring that adaptability is a core capability rather than an afterthought.

When a brand applies agility by design, it can quickly pivot strategies, products, and processes in response to market changes, technological advancements, and customer feedback. For example, a company might have modular product designs that allow for easy updates and customization or a flexible marketing strategy that can rapidly shift focus based on real-time data insights.

Understand That Change Is the Only Constant

Understanding that change is the only constant involves recognizing that the business environment is always evolving due to technological advancements, market trends, consumer behavior, and competitive dynamics. This mindset accepts that change is inevitable and views it as an opportunity for growth and innovation rather than a disruption.

Embracing the inevitability of change is foundational to Agility by Design. It ensures that the organization remains proactive, anticipating shifts and preparing to adapt quickly. By fostering a culture that views change positively, businesses can maintain their competitive edge and continuously improve their offerings and processes.

A tech company that understands change as a constant might regularly invest in research and development to stay ahead of industry trends. They continuously update their product lines to incorporate the latest technological advancements and address emerging customer needs. This proactive approach ensures they remain leaders in their field rather than being caught off-guard by market shifts.

Build to Change

Building to change means designing products, processes, and systems that are inherently flexible and adaptable. This approach involves creating modular and scalable solutions that can be easily updated or modified in response to new information or changing circumstances.

This concept directly supports Agility by Design by ensuring that the organization's infrastructure can accommodate and facilitate change. When systems are built with adaptability in mind, the business can pivot more easily, implement improvements swiftly, and integrate new technologies without extensive overhauls.

Consider a company that develops a modular software platform. Each component of the software can be updated or replaced independently without affecting the entire system. This design allows the company to quickly implement updates and new features based on user feedback and technological advancements, maintaining the software's relevance and effectiveness over time.

By incorporating these two points—understanding that change is the only constant and building to change—organizations can create a robust framework for agility. This foundation enables them to respond effectively to the dynamic nature of the market, ensuring sustained growth and innovation.

Common Roadblocks

1. **Rigid Organizational Structures:** Traditional hierarchical structures can impede quick decision-making and adaptability.
2. **Resistance to Change:** Employees and management might be resistant to changing established processes and systems.
3. **Lack of Resources:** Insufficient investment in tools, training, and technology can hinder the implementation of Agile practices.

How to Adopt It

To adopt agility by design, an organization must foster a culture that embraces change, supports continuous improvement, and encourages innovation. This involves creating flexible structures, empowering employees, and investing in the necessary resources and technologies.

5 Steps to Adoption of the Principle

1. **Evaluate Current Structures:** Assess existing organizational structures and processes to identify areas that are rigid and inflexible. Consider whether these structures hinder adaptability and where changes can be made to support agility.
2. **Promote a Culture of Change:** Cultivate a mindset that views change as an opportunity rather than a threat. This can be done through regular training sessions, workshops, and open

forums where employees can discuss and suggest improvements.

3. **Implement Modular Designs:** Design products, services, and processes in a modular way that allows for easy updates and customization. This approach enables quick adaptations to market demands and technological advancements.

4. **Invest in Agile Tools and Technologies:** Provide teams with the tools and technologies that support agile practices, such as project management software, collaboration platforms, and data analytics tools.

5. **Empower Teams:** Give teams the autonomy to make decisions and adapt processes as needed. Encourage cross-functional collaboration to ensure that different perspectives are considered in the decision-making process.

Conclusion

By following these steps, an organization can embed agility into its design, ensuring it remains responsive and adaptable in a rapidly changing business environment.

Are You There Yet? Answer These 3 Questions

In the Appendix, we provide a 7-part assessment for Agile Brand maturity. You can check your progress by rating your organization on a scale of 1-4 based on the questions below:

Principle 1: Agility by Design

Characteristic	Question
Flexible Processes	How adaptable are your processes to changing market demands?
Rapid Prototyping	How quickly can your team develop and test new ideas?
Dynamic Resource Allocation	How effectively can you reallocate resources to new priorities?

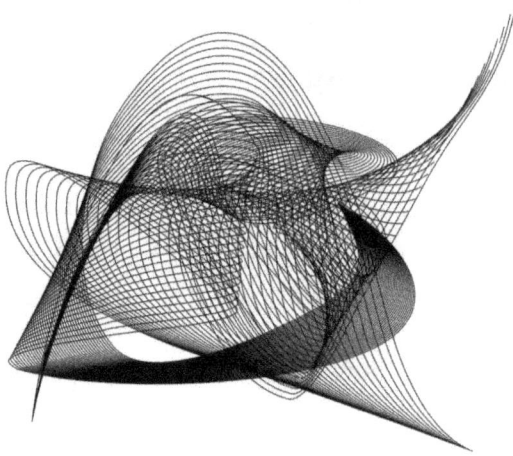

Chapter 5 |
Principle 2: Continuously Improving

Understanding when evolution vs. revolution is needed; Identify when evolutionary vs. revolutionary approaches are needed.

Continuous improvement is about fostering a culture of constant evaluation and enhancement within the organization. It involves regularly assessing processes, products, and strategies to identify areas for incremental improvements (evolution) and recognizing when more significant, transformative changes (revolution) are necessary. This principle ensures that the brand remains competitive, efficient, and aligned with market demands.

A brand that embraces continuous improvement will regularly update its offerings based on customer feedback, industry trends, and internal performance metrics. For instance, a software company might release frequent updates to fix bugs and add features (evolution) while also preparing for major version releases that overhaul the user interface and functionality (revolution). This brand is always in tune with the latest developments and proactively enhances its value proposition.

Understanding When Evolution vs. Revolution Is Needed

Understanding when evolution versus revolution is needed involves discerning whether a situation requires incremental improvements or a complete overhaul. Evolutionary changes are gradual and continuous, aimed at refining and optimizing existing processes, products, or services. Revolutionary changes, on the other hand, are radical transformations that fundamentally alter the organization's approach or offerings.

Recognizing the difference between these two types of changes is crucial for continuous improvement. By knowing when to apply each approach, an organization can ensure it is making the right adjustments at the right time. This discernment prevents unnecessary disruption while also avoiding stagnation, enabling the organization to stay competitive and responsive to market demands.

A retail company may implement evolutionary changes by optimizing its inventory management system to reduce waste and improve efficiency. However, if the entire retail landscape shifts to e-commerce, a revolutionary change may be necessary, such as overhauling their business model to focus on online sales and digital marketing strategies.

Identify When Evolutionary vs. Revolutionary Approaches Are Needed

Identifying when to use evolutionary versus revolutionary approaches requires a thorough analysis of the current situation, market conditions, and future trends. This involves assessing the impact of potential changes, understanding customer needs, and evaluating the organization's capacity for adaptation. Evolutionary changes are best suited for enhancing and optimizing, while revolutionary changes are necessary when fundamental shifts are required.

This ability to identify the appropriate approach ensures that the organization remains agile and effective in its continuous improvement efforts. By making informed decisions about which type of change to implement, a company can maximize its resources, minimize risks, and drive sustained growth and innovation.

An automotive manufacturer might use evolutionary changes to improve fuel efficiency and safety features in their existing car models. However, if there is a significant shift towards electric vehicles in the market, a revolutionary approach would be necessary, such as investing heavily in electric vehicle technology and rebranding the company to align with sustainable practices.

By understanding and identifying when to apply evolutionary versus revolutionary changes, organizations can effectively navigate the complexities of continuous improvement. This strategic approach allows them to remain competitive and responsive, ensuring long-term success and growth.

Common Roadblocks

1. **Complacency:** Organizations that become comfortable with their current success may resist the need for continuous improvement.
2. **Lack of Clear Metrics:** Without clear performance metrics and feedback mechanisms, identifying areas for improvement can be challenging.
3. **Fear of Change:** Employees and leaders may fear the uncertainty and disruption of both evolutionary and revolutionary changes.

How to Adopt It as an Organization

To adopt continuous improvement, an organization must instill a mindset of ongoing development and flexibility. This involves setting up systems to regularly gather feedback, analyze performance data, and encourage innovative thinking at all levels.

5 Steps to Adoption of the Principle

1. **Establish Clear Metrics:** Define key performance indicators (KPIs) that align with your business goals. Regularly track and analyze these metrics to identify areas that need improvement.

2. **Create Feedback Loops:** Implement mechanisms for gathering feedback from customers, employees, and other stakeholders. Use surveys, focus groups, and direct feedback channels to gain insights.

3. **Encourage a Growth Mindset:** Foster a culture where continuous learning and development are valued. Offer training programs and resources that promote skill enhancement and innovative thinking.

4. **Implement Agile Practices:** Adopt agile methodologies that support iterative improvements, such as regular review meetings, sprints, and retrospectives. These practices help teams quickly adapt and implement incremental changes.

5. **Recognize and Reward Innovation:** Encourage employees to propose improvements and recognize those who contribute innovative ideas. Create a safe environment where experimentation is welcomed and failures are seen as learning opportunities.

Conclusion

By embedding these steps into the organization's culture, brands can ensure they are continually improving, whether through small, incremental changes or significant, transformative shifts. This approach helps maintain relevance, competitiveness, and customer satisfaction.

Are You There Yet? Answer These 3 Questions

In the Appendix, we provide a 7-part assessment for Agile Brand maturity. You can check your progress by rating your organization on a scale of 1-4 based on the questions below:

Principle 2: Continuously Improving

Characteristic	Question
Regular Feedback Collection	How frequently do you collect feedback from customers and employees?
Incremental Enhancements	How often do you implement small, incremental improvements?
Performance Metrics Tracking	How consistently do you track and analyze performance metrics?

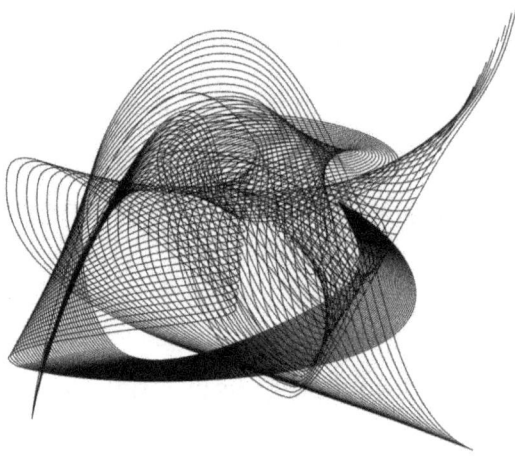

Chapter 6 |
Principle 3: Operationalizing adaptivity

Make adaptivity and collaboration part of your culture; embrace change, don't avoid it.

Operationalizing adaptivity involves embedding the ability to adapt and respond to change into the core operations of the organization. It's about creating an environment where adaptability and collaboration are not just encouraged but are integral parts of the daily workflow. This principle ensures that the organization remains flexible, responsive, and resilient in the face of changing market conditions and customer needs.

A brand that successfully operationalizes adaptivity will have systems and processes in place that allow for rapid adjustments to strategies, products, and operations. For example, a retail company might use real-time inventory management systems to quickly respond to changing demand patterns, or a marketing team might employ agile project management to rapidly pivot campaigns based on current trends. Collaboration tools and practices will be seamlessly integrated, enabling cross-functional teams to work together efficiently.

Make Adaptivity and Collaboration Part of Your Culture

Making adaptivity and collaboration part of your culture involves embedding these values into the organizational DNA. It means creating an environment where flexibility, teamwork, and open communication are encouraged and rewarded. Employees are empowered to share ideas, adapt to new information, and work together towards common goals.

This approach ensures that the organization can respond swiftly and effectively to changes. A culture of adaptivity and collaboration fosters innovation, enhances problem-solving capabilities, and enables the organization to pivot quickly in response to new challenges or opportunities. It creates a resilient and agile workforce better equipped to handle uncertainty and complexity.

A technology firm that prioritizes adaptivity and collaboration might implement cross-functional teams that work on projects together, blending diverse skills and perspectives. Regular brainstorming sessions and open forums for idea-sharing can lead to innovative solutions and rapid adaptations to market changes.

Embrace Change, Don't Avoid It

Embracing change means viewing change as an opportunity for growth and improvement rather than a threat. It involves actively seeking out and welcoming new ideas, technologies, and processes that can enhance the organization's performance and competitiveness. This mindset encourages experimentation and learning from failures as well as successes.

By embracing change, organizations can stay ahead of the curve and remain competitive in a rapidly evolving market. This proactive approach allows for continuous improvement and innovation, ensuring that the organization is not only reacting to changes but also driving them. It prevents stagnation and promotes a culture of continuous learning and development.

A retail company that embraces change might quickly adopt new technologies like AI-driven customer service tools or augmented reality shopping experiences. By doing so, they can enhance the customer experience, streamline operations, and stay competitive in a digital-first marketplace.

By making adaptivity and collaboration part of the culture and embracing change, organizations can operationalize adaptivity effectively. This approach ensures that the organization is always ready to respond to new challenges and opportunities, fostering a dynamic and resilient business environment.

Common Roadblocks

1. **Siloed Departments:** Lack of communication and collaboration between departments can hinder the ability to adapt quickly.
2. **Resistance to Change:** Employees and leadership may be resistant to changing established processes and embracing new ways of working.
3. **Inflexible Systems:** Legacy systems and rigid processes can prevent the organization from responding swiftly to changes.

How to Adopt Adaptivity in Your Organization

To adopt adaptivity, an organization must cultivate a culture that values the types of flexibility and collaboration we've been discussing in this book at length. As we've discussed, this involves encouraging open communication, breaking down silos, and investing in technologies and practices that support adaptive workflows.

5 Steps to Adoption

1. **Promote Cross-Functional Collaboration:** Encourage teams from different departments to work together on projects.

Collaboration tools like Slack, Microsoft Teams, or Trello can be used to facilitate communication and project management.

2. **Implement Agile Methodologies:** Adopt agile practices such as daily stand-ups, sprints, and retrospectives to create a more adaptive and responsive workflow. These practices help teams to quickly adjust their approaches based on feedback and new information.

3. **Train for Adaptivity:** Provide training and resources that help employees develop skills in adaptive thinking and agile methodologies. Encourage a mindset that sees change as an opportunity rather than a threat.

4. **Invest in Flexible Systems:** Upgrade or replace legacy systems with more flexible, scalable technologies that support quick changes and integrations. Cloud-based solutions and modular platforms can enhance adaptivity.

5. **Encourage Open Communication:** Foster an environment where employees feel comfortable sharing ideas, feedback, and concerns. Regularly hold meetings and forums where everyone can contribute to discussions about changes and improvements.

Conclusion

By implementing these steps, organizations can operationalize adaptivity, making it a natural and integral part of their culture and operations. This will enable them to respond swiftly and effectively to changes, maintaining their competitive edge and ensuring long-term success.

Are You There Yet? Answer These 3 Questions

In the Appendix, we provide a 7-part assessment for Agile Brand maturity. You can check your progress by rating your organization on a scale of 1-4 based on the questions below:

Principle 3: Operationalizing Adaptivity

Characteristic	Question
Cross-Functional Collaboration	How well do your teams collaborate across functions?
Scenario Planning	How prepared are you for potential market changes?
Adaptivity Training	How often do you provide training on adaptive methodologies?

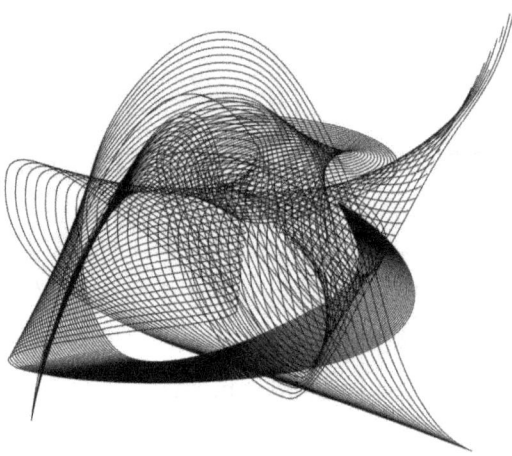

Chapter 7 |
Principle 4: Guided by values

Stay true to your values no matter what; respect customers and their data; respect employees and their time.

Being guided by values means that an organization's decisions and actions are consistently aligned with its core principles and ethical standards. This principle emphasizes the importance of integrity, transparency, and respect in all interactions with customers, employees, and stakeholders. It ensures that the brand's behavior reflects its proclaimed values, fostering trust and loyalty.

A brand that is guided by values will consistently demonstrate ethical behavior and decision-making. For example, such a brand will protect customer data, treat employees fairly, and uphold high standards of honesty and transparency. Patagonia, for instance, is known for its commitment to environmental sustainability, which is reflected in its product designs, corporate policies, and advocacy efforts. This alignment with values builds a strong, trust-based relationship with its audience.

Stay True to Your Values No Matter What

Staying true to your values means maintaining your core principles and ethical standards regardless of external pressures or circumstances. It involves making decisions that align with your organization's values, even when faced with challenges or opportunities that might tempt you to compromise. This steadfast commitment builds trust and credibility with customers, employees, and stakeholders.

Being guided by values provides a moral compass that ensures consistency and integrity in all actions and decisions. This consistency fosters trust and loyalty, as stakeholders can rely on the organization to act ethically and predictably. It also differentiates the brand, as customers and employees increasingly seek out organizations that stand for something meaningful and authentic.

Ben & Jerry's is known for its commitment to social justice and environmental sustainability. The company consistently advocates for these causes, even when it might not be the most profitable path. By staying true to its values, Ben & Jerry's has built a loyal customer base that supports its mission and trusts its brand.

Respect Customers and Their Data; Respect Employees and Their Time

Respecting customers and their data involves protecting their privacy and using their information ethically. It means being transparent about data practices and ensuring customer data is secure. Similarly, respecting employees and their time involves creating a work environment that values work-life balance, efficiency, and productivity. It means recognizing the importance of employee well-being and fostering a culture that supports it.

Respecting customers and their data builds trust and loyalty, as customers feel safe and valued. It also ensures compliance with data protection regulations, reducing legal and reputational risks. Respecting employees and their time leads to higher job satisfaction, better performance, and lower turnover rates. This mutual respect enhances overall organizational health and effectiveness.

Apple is known for its strong stance on customer privacy, often highlighting its commitment to protecting user data. This respect for customer data builds trust and sets the company apart in a tech industry where privacy concerns are rampant. Internally, companies like Netflix offer flexible work schedules and support remote work, respecting employees' time and promoting a healthy work-life balance.

By staying true to values and respecting customers and employees, organizations can create a strong ethical foundation that guides their actions and decisions. This approach not only enhances trust and loyalty but also ensures long-term sustainability and success.

Common Roadblocks

1. **Profit-Driven Decisions:** Pressure to meet short-term financial goals can lead to compromising on values.
2. **Inconsistent Leadership**: If leadership does not consistently model the organization's values, it can lead to a disconnect between stated values and actual practices.
3. **Lack of Accountability:** Without systems to hold individuals and teams accountable for upholding values, adherence can wane over time.

How to Adopt It as an Organization

To adopt a value-guided approach, an organization must embed its core values into every aspect of its operations and culture. This involves clear communication of values, consistent leadership, and mechanisms for accountability and transparency.

5 Steps to Adoption of the Principle

1. **Define and Communicate Core Values:** Clearly articulate the organization's core values and ensure they are communicated effectively to all employees. Use onboarding programs, internal communications, and regular training sessions to reinforce these values.
2. **Lead by Example:** Ensure that leaders and managers consistently demonstrate the organization's values in their decisions and actions. Leadership should model the behavior

they expect from their teams, setting a standard for integrity and ethical conduct.

3. **Implement Data Protection Policies:** Establish and enforce strict policies for protecting customer data. Ensure compliance with data protection regulations and educate employees about the importance of data privacy and security.

4. **Foster a Respectful Work Environment:** Create a culture of respect and inclusion for all employees. Implement policies that promote fair treatment, diversity, and employee well-being. Encourage open dialogue and provide channels for employees to voice concerns.

5. **Establish Accountability Mechanisms:** Develop systems to hold individuals and teams accountable for upholding the organization's values. This can include performance reviews, ethical audits, and feedback mechanisms that assess alignment with core values.

Conclusion

By following these steps, an organization can ensure that its actions are consistently guided by its values, fostering a culture of integrity, respect, and trust. This alignment will strengthen relationships with customers and employees, enhancing the brand's reputation and long-term success.

Are You There Yet? Answer These 3 Questions

In the Appendix, we provide a 7-part assessment for Agile Brand maturity. You can check your progress by rating your organization on a scale of 1-4 based on the questions below:

Principle 4: Guided by Values

Characteristic	Question
Ethical Decision-Making	How consistently do you ensure decisions align with core values?
Transparency	How openly do you communicate business decisions and changes?
Value-Driven Leadership	How well do leaders exemplify the organization's values?

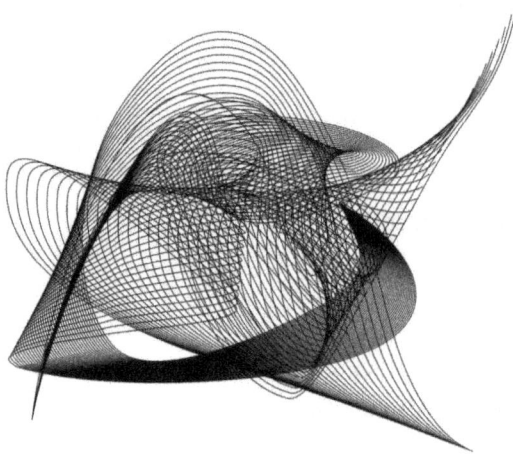

Chapter 8 |
Principle 5: Building relationships

Focus on the customer relationship, not the individual transactions. It's not about channels and mediums; it's about customer experience.

Building relationships is about prioritizing long-term customer connections over individual transactions. This principle emphasizes the importance of understanding and enhancing the overall customer experience across all touchpoints rather than focusing solely on sales or specific channels. It's about creating meaningful interactions that foster loyalty and trust.

A brand that excels in building relationships will have a customer-centric approach that permeates every aspect of its operations. For example, Zappos is renowned for its exceptional customer service, which focuses on building lasting relationships rather than just completing sales. This approach includes personalized service, hassle-free returns, and a commitment to exceeding customer expectations. The brand's emphasis on customer experience ensures that every interaction strengthens the relationship.

Focus on the Customer Relationship, Not the Individual Transactions

Focusing on the customer relationship means prioritizing long-term engagement and loyalty over short-term sales. This approach involves understanding and meeting the broader needs and preferences of customers rather than just completing individual transactions. It's about creating value and trust that extends beyond a single purchase.

By focusing on the relationship, organizations can build a loyal customer base that provides repeat business and referrals. This long-term perspective encourages companies to deliver consistent, high-quality service and to engage customers in meaningful ways. It leads to higher customer lifetime value and stronger brand loyalty.

Amazon's Prime membership program is a great example of focusing on the customer relationship. Instead of just facilitating individual purchases, Amazon provides a suite of services (free shipping, streaming, exclusive deals) that enhance the overall customer experience and foster long-term loyalty.

It's Not About Channels and Mediums; It's About Customer Experience

Prioritizing customer experience over channels and mediums involves creating seamless and consistent interactions across all touchpoints. It means ensuring that whether a customer interacts with the brand online, in-store, via social media, or through customer service, the experience is cohesive and aligned with the brand's values and promises.

Focusing on the customer experience ensures customers feel valued and understood at every interaction. It reduces friction, enhances satisfaction, and fosters loyalty. By viewing all touchpoints as part of a unified experience, organizations can deliver more personalized and effective service, improving overall customer satisfaction and retention.

Starbucks excels at creating a consistent customer experience across all channels. Whether customers visit a physical store, use the mobile app, or interact with Starbucks on social media, they encounter a cohesive brand experience. This integrated approach enhances convenience and satisfaction, reinforcing customer loyalty.

By focusing on the customer relationship and prioritizing customer experience, organizations can build strong, lasting connections with their customers. This approach ensures that every interaction adds value and strengthens the overall relationship, driving long-term success and loyalty.

Common Roadblocks

1. **Short-Term Focus:** Companies often prioritize immediate sales and revenue over long-term relationship building, which can undermine customer trust and loyalty.
2. **Channel Siloes:** Treating channels and mediums as separate entities can lead to inconsistent customer experiences and fragmented communication.
3. **Lack of Customer Insight:** Without a deep understanding of customer needs and preferences, it's challenging to create meaningful and personalized experiences.

How to Adopt It

To adopt a relationship-building approach, an organization must shift its focus from individual transactions to the overall customer journey. This involves integrating customer insights, ensuring consistent experiences across all channels, and fostering a culture prioritizing customer satisfaction and loyalty.

5 Steps to Adoption of the Principle

1. **Map the Customer Journey:** Understand and document the entire customer journey, identifying key touchpoints and pain points. Use this map to ensure that every interaction enhances the overall experience.

2. **Integrate Customer Data:** Use a centralized customer relationship management (CRM) system to collect and analyze customer data from all channels. This integration helps create a unified view of the customer and enables personalized interactions.

3. **Train Employees on Customer-Centric Practices:** Provide training for employees at all levels to emphasize the importance of relationship building and customer experience. Encourage empathy, active listening, and proactive problem-solving in customer interactions.

4. **Ensure Consistency Across Channels:** Develop a strategy that ensures a seamless and consistent customer experience across all channels and touchpoints. Align messaging, service standards, and branding to create a cohesive journey.

5. **Gather and Act on Customer Feedback:** Implement systems to regularly collect feedback from customers and use it to improve products, services, and interactions. Show customers that their input is valued and acted upon to build trust and loyalty.

Conclusion

By focusing on these steps, organizations can build strong, lasting relationships with their customers, enhancing loyalty and creating advocates for their brand. This approach ensures that every interaction contributes to a positive and cohesive customer experience, ultimately driving long-term success.

Are You There Yet? Answer These 3 Questions

In the Appendix, we provide a 7-part assessment for Agile Brand maturity. You can check your progress by rating your organization on a scale of 1-4 based on the questions below:

Principle 5: Building Relationships

Characteristic	Question
Customer Engagement	How effectively do you engage with customers beyond transactions?
Employee Engagement	How well do you engage with employees and foster their loyalty?

Relationship Metrics	How regularly do you measure the quality of your relationships?

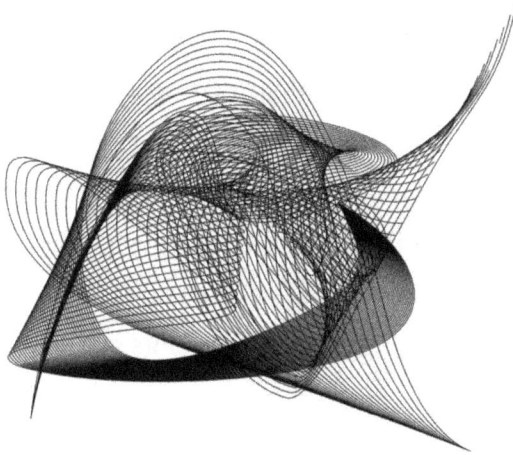

Chapter 9 |
Principle 6: Focusing on the
conversation

Listen to your customers and your employees; don't talk at them. Love the dialogue; there is always something to learn from it.

Focusing on the conversation means actively engaging in two-way communication with both customers and employees. Rather than delivering one-sided messages, this principle emphasizes the importance of listening and valuing dialogue. It recognizes that meaningful conversations can provide valuable insights, foster trust, and drive continuous improvement.

A brand that prioritizes conversation will create numerous channels for open dialogue and actively seek out feedback. For example, Buffer, a social media management company, practices radical transparency and open communication with its customers and employees. They regularly share company updates, seek input through surveys and social media, and encourage open discussions. This approach not only builds trust but also ensures that the brand remains aligned with the needs and expectations of its stakeholders.

Listen to Your Customers and Your Employees, Don't Talk at Them

Listening to your customers and employees means actively seeking their input, understanding their needs and concerns, and valuing their perspectives. It involves creating opportunities for them to share feedback and ensuring that their voices are heard and considered in decision-making processes. This approach fosters a sense of engagement and respect.

By listening rather than just talking, organizations can build deeper, more meaningful relationships. This approach helps identify pain points, opportunities for improvement, and areas where the organization can better meet the needs of its stakeholders. It leads to increased trust, loyalty, and a stronger sense of community within the organization and with its customers.

Salesforce regularly surveys its customers and employees to gather feedback on their products and workplace environment. By actively listening and making adjustments based on this feedback,

Salesforce demonstrates that it values the input of its stakeholders, leading to higher satisfaction and engagement levels.

Love the Dialogue, There Is Always Something to Learn from It

Loving the dialogue involves embracing open communication and seeing every conversation as an opportunity to learn and grow. It means valuing feedback, questions, and discussions and approaching them with curiosity and a willingness to improve. This mindset fosters continuous learning and adaptation.

Embracing dialogue encourages a culture of transparency and continuous improvement. By valuing every conversation, organizations can gain valuable insights, identify trends, and make informed decisions. This approach promotes innovation and agility, as the organization is constantly learning from its interactions with customers and employees.

Zappos is renowned for its customer service, which emphasizes genuine conversations with customers. Customer service representatives are encouraged to spend as much time as needed on calls, fostering authentic interactions and gathering insights to improve the customer experience. This love for dialogue helps Zappos continuously enhance its service and build strong customer relationships.

By listening to customers and employees and valuing dialogue, organizations can foster an environment of mutual respect and continuous learning. This approach ensures that the organization remains responsive, innovative, and aligned with the needs of its stakeholders, driving long-term success and engagement.

Common Roadblocks

1. 1. One-Way Communication: Traditional marketing and management approaches often focus on broadcasting messages rather than engaging in dialogue.
2. 2. Ignoring Feedback: Collecting feedback without acting on it can lead to frustration and disengagement among customers and employees.
3. 3. Fear of Criticism: Concerns about receiving negative feedback can prevent organizations from genuinely engaging in conversations.

How to Adopt It

To adopt a conversation-focused approach, an organization must cultivate a culture that values open communication and active listening. This involves creating platforms for dialogue, training employees on effective communication skills, and demonstrating a commitment to acting on the feedback received.

5 Steps to Adoption

1. **Create Multiple Feedback Channels:** Establish various platforms for customers and employees to share their thoughts, such as surveys, social media, forums, and suggestion

boxes. Ensure these channels are easily accessible and user-friendly.

2. **Train on Active Listening:** Provide training for employees at all levels on active listening techniques. Encourage them to engage fully in conversations, show empathy, and seek to understand rather than simply respond.

3. **Act on Feedback:** Develop a systematic approach to review, analyze, and act on the feedback received. Communicate the actions taken in response to feedback to show that it is valued and has an impact.

4. **Encourage Open Dialogue:** Foster a culture where open dialogue is encouraged and valued. Hold regular meetings, town halls, and open forums where employees and customers can voice their opinions and ideas.

5. **Celebrate the Conversation:** Highlight examples where feedback has led to positive changes. Share stories of successful dialogues to demonstrate the value of conversations and encourage ongoing participation.

Conclusion

By embedding these steps into their practices, organizations can foster meaningful conversations with their customers and employees. This approach ensures that the brand remains responsive, innovative, and aligned with the needs of its stakeholders, ultimately driving long-term success.

Are You There Yet? Answer These 3 Questions

In the Appendix, we provide a 7-part assessment for Agile Brand maturity. You can check your progress by rating your organization on a scale of 1-4 based on the questions below:

Principle 6: Focusing on the Conversation

Characteristic	Question
Active Listening	How effectively do you listen to and act on stakeholder feedback?
Open Dialogue	How frequently do you facilitate open discussions with stakeholders?
Feedback Implementation	How well do you implement changes based on feedback received?

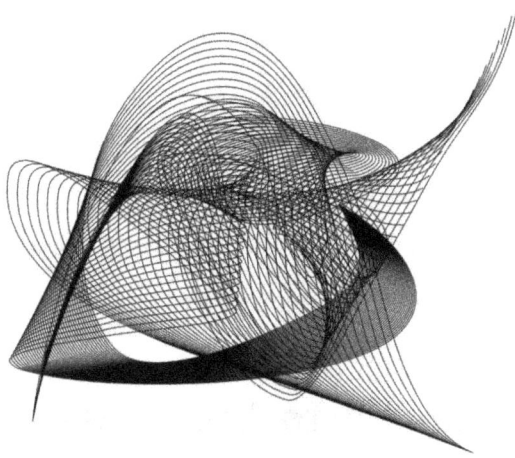

Chapter 10 |
Principle 7: Always learning and growing

Have the humility to acknowledge room for improvement; Create opportunities for better ideas; Continuously learn. Continuously improve.

Always learning and growing is about fostering a mindset of continuous improvement and development within an organization. This principle emphasizes humility, openness to new ideas, and a commitment to ongoing education. It acknowledges that there is always room for improvement and that growth comes from embracing new knowledge and innovations.

A brand that embraces continuous learning will have systems and cultures that encourage ongoing education, experimentation, and feedback. For example, Google is known for its "20% time" policy, which allows employees to spend 20% of their time working on projects they are passionate about, leading to innovations like Gmail and Google Maps. This approach nurtures creativity, drives innovation, and ensures the company always evolves.

Have the Humility to Acknowledge Room for Improvement

Having the humility to acknowledge room for improvement means recognizing that no matter how successful or efficient an organization is, there are always areas that can be enhanced. This attitude requires an openness to feedback, a willingness to admit mistakes, and a commitment to personal and organizational growth.

Humility fosters a culture of continuous improvement by encouraging everyone in the organization to seek out and act on opportunities for enhancement. It creates an environment where employees feel safe to share honest feedback and innovative ideas without fear of judgment. This approach leads to incremental and sometimes transformative improvements that keep the organization competitive and responsive.

Toyota's commitment to Kaizen, or continuous improvement, embodies this principle. Employees at all levels are encouraged to suggest improvements, and the company continually seeks ways to enhance processes and products. This humility and openness to change have been key to Toyota's long-term success and innovation.

Create Opportunities for Better Ideas

Creating opportunities for better ideas involves fostering an environment where creativity and innovation are encouraged and supported. This means providing employees the tools, resources, and platforms to brainstorm, share, and develop new ideas. It also involves recognizing and rewarding innovative thinking.

By actively seeking and nurturing better ideas, organizations can drive innovation and stay ahead of the competition. This proactive approach ensures a steady flow of fresh perspectives and solutions that can enhance products, services, and processes. It also engages employees by making them feel valued and empowered to contribute to the organization's success.

Google's "20% time" policy allows employees to spend 20% of their work time on projects they are passionate about. This policy has led to the development of major products like Gmail and Google News, demonstrating how creating opportunities for innovation can lead to significant advancements and new revenue streams.

Continuously Learn. Continuously Improve.

Continuously learning and improving means adopting a mindset of perpetual education and development. It involves staying curious, seeking out new knowledge, and applying what is learned to make ongoing enhancements. This principle encourages organizations to remain adaptable and forward-thinking.

Continuous learning and improvement ensure that an organization remains dynamic and competitive. By constantly seeking to improve, organizations can better meet their customers' evolving needs and adapt to market changes. This approach leads to sustained growth and innovation.

Amazon's culture of continuous improvement is reflected in its commitment to innovation and customer focus. The company continually iterates on its processes and technologies to enhance customer experience, from its recommendation algorithms to its delivery logistics. This relentless pursuit of improvement has helped Amazon maintain its position as a market leader.

Embracing the principles of humility, innovation, and continuous learning is essential for any organization aiming to thrive in a competitive and ever-changing environment. Organizations can drive sustained growth, innovation, and success by acknowledging room for improvement, creating opportunities for better ideas, and fostering a culture of continuous learning and improvement.

Common Roadblocks

1. **Complacency:** Success can breed complacency, leading to resistance to change and a lack of drive for continuous improvement.

2. **Fear of Failure:** A culture that penalizes failure can stifle innovation and discourage employees from experimenting or proposing new ideas.

3. **Lack of Resources:** Insufficient time, budget, or support for learning and development initiatives can hinder continuous growth.

How to Adopt It

To adopt a culture of continuous learning and growth, an organization must create an environment that encourages experimentation, values feedback, and provides resources for ongoing education. This involves setting up formal and informal learning opportunities, fostering a growth mindset, and celebrating innovation.

5 Steps to Adoption

1. **Establish a Learning Culture:** Promote the importance of continuous learning and development throughout the organization. Encourage employees to seek out new knowledge and skills regularly.

2. **Provide Learning Opportunities:** Offer a variety of learning and development programs, such as workshops, online courses, mentorship, and conferences. Ensure that employees have access to resources that support their growth.

3. **Encourage Experimentation:** Create a safe environment where employees feel comfortable

experimenting with new ideas and approaches. Emphasize that failures are learning opportunities and integral to innovation.

4. **Solicit and Act on Feedback:** Regularly seek feedback from employees and customers to identify areas for improvement. Implement changes based on this feedback to demonstrate that it is valued and acted upon.

5. **Celebrate Growth and Innovation:** Recognize and reward employees who contribute innovative ideas and show significant personal or professional growth. Share success stories to inspire others and reinforce the importance of continuous learning.

Conclusion

By embedding these steps into their organizational practices, companies can cultivate a culture of continuous learning and growth. This approach ensures that the brand remains innovative, adaptable, and competitive in an ever-evolving market, driving long-term success and sustainability.

Are You There Yet? Answer These 3 Questions

In the Appendix, we provide a 7-part assessment for Agile Brand maturity. You can check your progress by rating your organization on a scale of 1-4 based on the questions below:

Principle 7: Always Learning and Growing

Characteristic	Question
Continuous Learning Programs	How regularly do you offer learning and development opportunities?
Innovation Encouragement	How effectively do you encourage and support innovation?
Knowledge Sharing	How well do you facilitate the sharing of knowledge across the organization?

This concludes our exploration of the seven principles of the Agile Brand, and while it may seem like a lot to undertake to embody all of the principles, I assure you it is possible. To back that up, we will look at some brands you're likely familiar with already and see how they apply the concepts and principles we've just reviewed to their organizations. You will likely notice that, even though they embrace the same principles, they undertake them in different ways.

Now, we will look at three companies that incorporate these seven principles into their DNA.

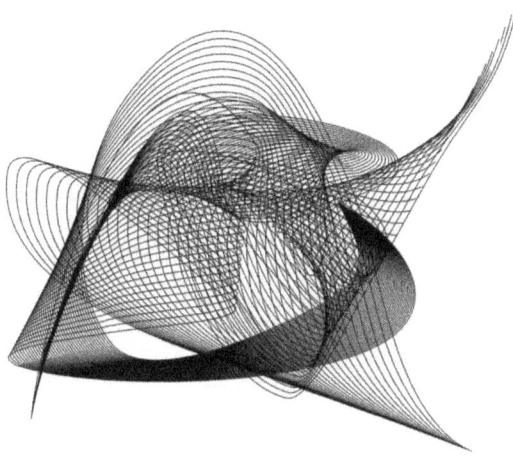

Chapter 11 |
An Agile Brand Case Study:
Spotify

Spotify, the world's leading music streaming service, has revolutionized how people listen to and discover music. Founded in 2006, Spotify has grown exponentially, now boasting over 615 million active users worldwide[9]. This case study examines how Spotify embodies the principles of an Agile Brand, maintaining its competitive edge and fostering strong relationships with both users and artists.

Principle 1: Agility by Design

Spotify's organizational structure is designed for agility, utilizing a model that includes squads, tribes, chapters, and guilds. This setup allows for rapid prototyping and deployment of new features, ensuring the company can quickly adapt to changing market conditions and user preferences. For instance, the development and success of Discover Weekly, a personalized playlist feature, exemplifies how Spotify leverages agility to create innovative and user-centric products[10].

Principle 2: Continuously Improving

Spotify's commitment to continuous improvement is evident in its iterative development processes. The company regularly collects and integrates user feedback into its product development cycle, ensuring that updates and new features meet user needs. This approach is reflected in the continuous updates to Spotify's user interface and algorithm enhancements designed to improve user experience and music discovery[11].

Principle 3: Operationalizing Adaptivity

Operationalizing adaptivity is a core part of Spotify's strategy. The company's decision-making processes are highly adaptive, allowing it to respond swiftly to market trends and user preferences. A prime example is Spotify's integration of podcasts into its platform, which was a strategic move to adapt to the growing demand for podcast content and diversify its offerings[12].

Principle 4: Guided by Values

Spotify's core values—innovation, passion, collaboration, and sincerity—guide its business decisions and operations. The company's commitment to these values is evident in initiatives like Spotify for Artists, which provides transparency and resources to artists about their streaming data and earnings, fostering a fairer and more transparent music industry.

Principle 5: Building Relationships

Building and maintaining strong relationships with users and artists is central to Spotify's strategy. The platform's personalized playlists and artist promotion features demonstrate its focus on engagement and value creation. By offering tailored music recommendations and supporting artist visibility, Spotify strengthens its relationships with key stakeholders.

Principle 6: Focusing on the Conversation

Spotify actively engages in meaningful dialogue with both users and artists through various channels. Social media engagement and in-app feedback mechanisms allow Spotify to effectively listen to and act on stakeholder feedback. This ongoing conversation helps the company stay attuned to the needs and preferences of its community.

Principle 7: Always Learning and Growing

Spotify fosters a culture of continuous learning and development through training programs, workshops, and innovation labs. The company regularly hosts hackathons to encourage creativity and innovation among its employees. These initiatives ensure that Spotify remains at the forefront of industry trends and technological advancements.

Conclusion

Spotify exemplifies the principles of an Agile Brand through its commitment to agility, continuous improvement, adaptivity, value-driven operations, relationship building, engaging conversations, and perpetual learning. These principles have enabled Spotify to maintain its market leadership and build a successful, sustainable brand. The case of Spotify highlights the critical role of agility in navigating a rapidly evolving market and achieving long-term success.

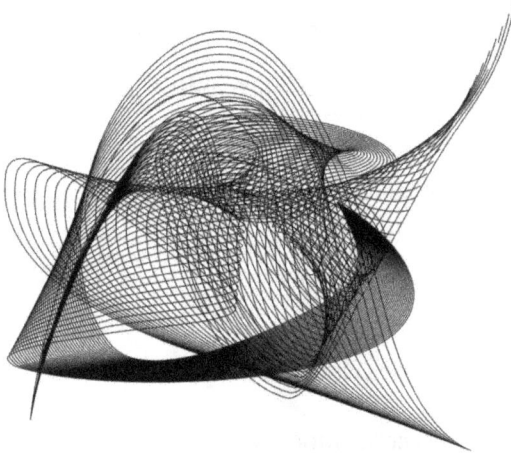

Chapter 12 |
An Agile Brand Case Study: Nike

Nike, founded in 1964, has grown to become a global leader in athletic footwear, apparel, and equipment[13]. Known for its innovative approach and commitment to excellence, Nike exemplifies the principles of an Agile Brand. This case study explores how Nike embodies these principles through its adaptive strategies, customer-centric focus, and continuous improvement.

Principle 1: Agility by Design

Nike's organizational structure is designed for agility, utilizing a matrix structure that promotes flexibility and rapid response to market changes. This structure allows for efficient decision-making and collaboration across different functions and regions. For example, during the COVID-19 pandemic, Nike swiftly adapted its manufacturing processes to produce personal protective equipment (PPE), showcasing its ability to pivot quickly in response to urgent needs[14].

Principle 2: Continuously Improving

Nike's commitment to continuous improvement is evident in its iterative development processes and sustainability efforts. The company regularly integrates athlete and customer feedback into its product development cycles, ensuring its offerings meet and exceed expectations. A notable example is the development of Nike Flyknit technology, which revolutionized athletic footwear by reducing waste and enhancing performance[15].

Principle 3: Operationalizing Adaptivity

Nike's operations are designed to be highly adaptive, allowing the company to quickly adjust to market demands and consumer preferences. This adaptability is exemplified by Nike's ability to shift production and supply chain strategies in response to changing conditions. The company's supply chain hack, which involves real-time data and flexible logistics, enables Nike to stay ahead of market trends and maintain operational efficiency.

Principle 4: Guided by Values

Nike's core values of diversity, inclusion, and sustainability guide its business decisions and operations. The company's Move to Zero campaign, aimed at reducing carbon emissions and waste, reflects Nike's commitment to environmental responsibility[16]. Additionally, initiatives like Nike's diversity and inclusion programs ensure that the company fosters an inclusive workplace and supports social justice.

Principle 5: Building Relationships

Building strong relationships with customers and athletes is central to Nike's strategy. The company engages its audience through personalized experiences and community-driven initiatives. The Nike Training Club app, which offers customized workout plans and connects users with professional trainers, is a prime example of how Nike fosters engagement and builds loyalty among its customers.

Principle 6: Focusing on the Conversation

Nike actively engages in meaningful dialogue with its customers and athletes through various channels, including social media and in-app feedback mechanisms. This two-way communication allows Nike to stay attuned to consumer needs and preferences, enabling the company to make informed decisions. Nike's social media campaigns, which encourage user participation and feedback, exemplify this approach.

Principle 7: Always Learning and Growing

Nike fosters a culture of continuous learning and innovation through training programs, workshops, and innovation labs. The Nike Innovation Kitchen, where new product ideas are developed and tested, highlights the company's commitment to fostering creativity and staying at the forefront of technological advancements[17]. These initiatives ensure that Nike remains a leader in the industry by continually enhancing its products and processes.

Conclusion

Nike exemplifies the principles of an Agile Brand through its agility, continuous improvement, adaptivity, value-driven operations, relationship building, engaging conversations, and perpetual learning. These principles have enabled Nike to maintain its market leadership and build a successful, sustainable brand. The case of Nike highlights the critical role of agility in navigating a rapidly evolving market and achieving long-term success.

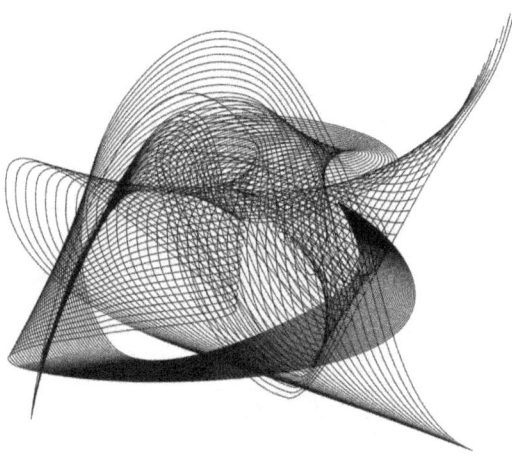

Chapter 13 |
An Agile Brand Case Study: Netflix

Netflix, founded in 1997, has transformed from a DVD rental service into the world's leading streaming platform, fundamentally changing how audiences consume media. Known for its innovative approach and commitment to user satisfaction, Netflix epitomizes the principles of an Agile Brand. This case study explores how Netflix embodies these principles through adaptive strategies, continuous improvement, and a strong emphasis on customer experience.

Principle 1: Agility by Design

Netflix's organizational structure promotes agility through a culture of freedom and responsibility, empowering employees to make decisions and innovate rapidly[18]. This design has allowed Netflix to pivot from DVD rentals to streaming services and, later, to original content production, demonstrating its ability to adapt quickly to market changes and technological advancements.

Example

The shift from DVD rentals to streaming and then to original content like "House of Cards" and "Stranger Things" showcases Netflix's ability to adapt its business model in response to changing consumer preferences and technological advancements[19].

Principle 2: Continuously Improving

Netflix's commitment to continuous improvement is evident in its data-driven decision-making and iterative development processes. The company uses extensive data analytics to understand viewer preferences and improve its content offerings. Netflix regularly tests new features and updates its platform to enhance user experience.

Example

A good example of continuous improvement is the development and ongoing refinement of Netflix's recommendation algorithm, which provides personalized content suggestions based on user viewing data. This algorithm has famously gone through several iterations over the years, including its offer in 2006 of $1 million for the team that developed a formula that improved the accuracy of their recommendations[20].

Principle 3: Operationalizing Adaptivity

To respond to regional preferences and trends, Netflix operationalizes adaptivity by leveraging its global presence and diverse content library. The company's decision-making processes are highly adaptive, allowing it to quickly pivot in response to market demands.

Example

Netflix's strategy to produce and acquire local content, such as "Money Heist" in Spain and "Sacred Games" in India, to cater to regional audiences and preferences allows it to stay nimble and offer content tailored to specific audiences and their preferences.

Principle 4: Guided by Values

Netflix's core values of innovation, customer obsession, and inclusivity guide its operations and decisions. The company's culture emphasizes honesty, transparency, and responsibility, ensuring that all actions align with these values.

Example

Netflix's open work culture, where employees are encouraged to speak their minds and are empowered to make impactful decisions, reflects its commitment to transparency and inclusivity. This is available in a publicly available statement available on their website[21].

Principle 5: Building Relationships

Building strong relationships with customers and content creators is central to Netflix's strategy. The company engages its audience through personalized experiences, fostering strong connections with creators.

Examples

Netflix's investment in original content and partnerships with top-tier talent to create exclusive shows and movies that resonate with its audience are a great example of relationships at work.

Customer lifetime value (CLV) also plays a strong role in strategy at Netflix, and some researchers from the University of Montana calculated the lifetime value of a customer to be $836.83 USD as of 2024[22]. Knowing this and embracing the lifetime value model helps Netflix stay focused on building long-term, sustainable relationships based on providing consistently good content and great experiences.

Principle 6: Focusing on the Conversation

Netflix actively engages in meaningful dialogue with its customers through various channels, including social media and in-app feedback mechanisms. This two-way communication allows Netflix to stay attuned to consumer needs and preferences, enabling the company to make informed decisions.

Example

Netflix has a robust social media presence, where it interacts with fans, responds to feedback, and fosters community engagement around its content[23].

Principle 7: Always Learning and Growing

Netflix fosters a culture of continuous learning and development through regular training programs, workshops, and innovation labs. The company encourages employees to experiment and innovate, ensuring it remains at the forefront of the entertainment industry.

Example

The establishment of Netflix's Post-Production Technology Department allows the company to focus on developing new tools and technologies to enhance content creation and distribution.

Conclusion

Netflix exemplifies the principles of an Agile Brand through its agility, continuous improvement, adaptivity, value-driven operations, relationship building, engaging conversations, and perpetual learning. These principles have enabled Netflix to maintain its market leadership and build a successful, sustainable brand. The case of Netflix highlights the critical role of agility in navigating a rapidly evolving market and achieving long-term success.

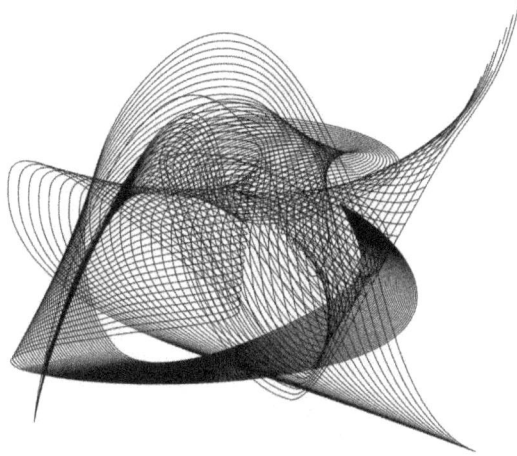

Part 3 |
Building the Agile Brand

"It takes 20 years to build a reputation and five
minutes to ruin it. If you think about that, you'll do
things differently." —Warren Buffett

We've explored the foundations of branding, and in particular, those of an Agile Brand. By exploring the principles in the last section, we have a good understanding of what guides an Agile Brand, and by now, you should understand that this is always a work in progress. For better or worse, the work of creating an Agile Brand is never done.

In this last section of the book, we will look at the elements that go into building and maintaining an Agile Brand. After all, it is not enough to start with the best intentions. Because the world changes rapidly around us—customers, competition, employees, and the world at large—maintaining agility means finding ways to continually grow and continually improve.

Let's look at how this is done.

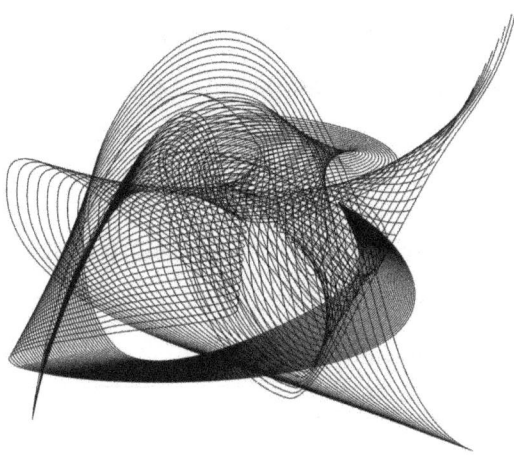

Chapter 14 |
The Responsibilities of an
Agile Brand

TL;DR

An Agile Brand has a *responsibility* to its customers to provide a rewarding experience that matches or surpasses the quality of its products and services.

An Agile Brand has a *responsibility* to its employees to provide a great experience, plus the opportunity for *everyone* to be part of the ideas and growth of the organization.

An Agile Brand has a *responsibility* to its partners and vendors to create ethical and sustainable relationships while promoting diversity.

An Agile Brand has a *responsibility* to the future to make the world a better place in the ways that it can influence and improve it.

It is important to address one other key aspect of brands that is part of the "new normal." It's not enough to make shareholders profits. It's also not enough to simply make a great product or offer an amazing service to customers. Additionally, it's also not enough to simply engage with our audiences on a regular basis. There needs to be something more, and truly successful brands have tapped into this.

Just as an Agile Brand must establish a bi-directional relationship with individual consumers to be successful, there must also be a symbiotic relationship with the world at large. This can play out in ways that impact society, the environment, or other areas that have a large cultural impact.

A company has many responsibilities to a variety of stakeholders. In addition to its customers and employees (including leadership), there are board members, shareholders, partners, the public at large, and potentially many other niche audiences. While some of these responsibilities are financial, and others may be to uphold a level of service, there is yet another category that is increasingly discussed and important to the long-term success of an organization.

The brand of today and tomorrow must be an Agile one, nimble enough to adapt to fast-changing environments and rooted in continually improving the people, processes, and systems that comprise it. In this chapter, I'm going to talk about the responsibilities of the Agile, modern brand that go *beyond* financial and customer service ones.

Customers

Let's start with that critical external audience: customers. While a brand is responsible for delivering a great product or service, both initially and well after the sale, we're not here to talk about that. Instead, an Agile Brand has a responsibility to provide the best possible experience before, during, and after the sale.

A good experience is not a "nice to have" feature anymore, either. According to Salesforce, 88% of customers believe personalized experiences are as important as a brand's products/services[24]. If you think about that for a minute, that is quite a statistic!

Thus, an Agile Brand has a *responsibility* to its customers to provide a rewarding experience that matches or surpasses the quality of its products and services. Customers increasingly expect this, and they vote with their wallets when it comes to choosing which brands provide the best experience.

Additionally, it's not solely about *experience*. The modern brand also has a responsibility to its audiences—customers chief among them—to express and make clear its values. Note that this doesn't mean a brand's values have to match any one group of individuals or majority, but there is a growing responsibility that the brand makes it clear what it stands for and its code of ethics. We saw an increased focus on this with the influx of the Millennial generation into both the workforce and increased buying power as consumers. However, note that this is not simply relegated to a single generation.

Employees

While they are sometimes not treated as such, the next audience can be as important or even more important than the first. The employee experience has run into challenges due to rapid and rather massive shifts in the employment numbers as well as the pandemic and its impact on in-person, remote, and hybrid work. The modern, Agile Brand has a responsibility to its employees to provide them with clarity, respect, and consistency in what the expectations are in regard to what their employee experience should consist of. Unfortunately, there is no one-size-fits-all approach here.

Also, like my comments above about brands' commitment to articulating and sharing their values with *customers*, employees also want to work somewhere that aligns with their values. With growing DEI programs as well as ESG efforts, brands are responding here, but there is often work to be done to pay more than lip service to these efforts. Employees and customers alike are continuing to grow more savvy in their efforts to tell when efforts to establish values are simply words, not actions.

To underscore this, a brand's ethics, as well as its commitment to diversity and inclusion, make for a stronger company overall. This is the win-win component: greater diversity and inclusion of ideas, viewpoints, and *people* leads to better strategies, more creativity, and a more sustainable organization.

Thus, an Agile B rand has a *responsibility* to its employees to provide a great experience, plus the opportunity for *everyone* to be part of the ideas and growth of the organization.

Partners and Vendors

Similar to how both customers and employees are treated, partners and vendors should also receive a good experience and be treated ethically. After all, with some of the processes and systems that create increased interchangeability that customers benefit from comes the ability for partners and vendors to choose who they partner with more easily as well.

Additionally, diversity should be sought when seeking partnerships and vendor relationships for reasons similar to those for employees: more diverse ideas and perspectives bring better results.

Thus, an Agile Brand has a *responsibility* to its partners and vendors to create ethical and sustainable relationships while promoting diversity.

The world at large and the future

Does a brand owe anything to those who aren't stakeholders, customers, or employees? I would argue that, yes, it absolutely does. A company's responsibility to the world at large can vary, depending on what exact impact it has on energy, the environment, privacy, quality of life, and countless other factors.

While I won't get specific here due to the wide range of potential impacts, suffice it to say a brand should be thinking about its reputation and impact beyond its immediate stakeholders like customers, employees, and shareholders.

So, what about a brand's responsibility beyond the current point in time and maybe the immediate future? A brand should lay a foundation for the future. This means taking a long-term view of its customers, employees, and the world at large.

Thus, an Agile Brand has a *responsibility* to the future to make the world a better place in the ways it can influence and improve it.

As you can see, creating and sustaining an Agile Brand includes a multitude of considerations, including the responsibility of the brand to live up to its values and ethics, as well as to invest in the sustainability of the organization itself and the world at large.

Conclusion

Social responsibility must be more than a marketing ploy to reach younger generations because it's something they care about. Instead, it must be translated into everything the company does. Consumers in general (not just younger generations) want to feel good about the choices they make and the dollars they spend.

The Agile Brand understands this and incorporates this thinking throughout everything it does instead of simply choosing words that sound good on its marketing materials. Social responsibility, clear communication of ethics, and a commitment to diversity, equity, and inclusion are becoming a prerequisite for being competitive as a business. Agile Brands understand that their values matter and must be genuinely expressed and demonstrated.

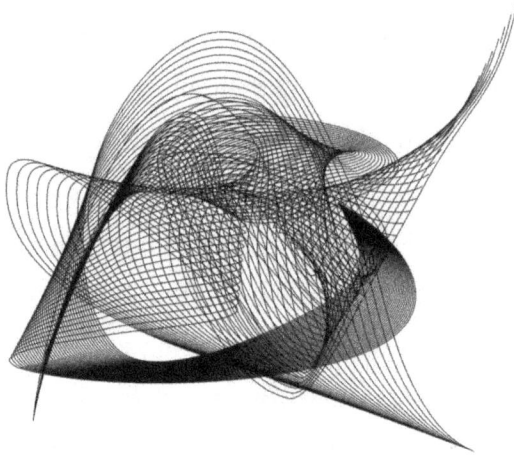

Chapter 15 |
The Duality of the Agile Brand

TL;DR

An Agile Brand can have elements that change and adapt over time, yet those are balanced by other more timeless elements like values and mission. These elements work together to form the duality of the Agile Brand.

Once, when I was speaking on the topic of Agile marketing at a conference, someone asked a question that should be addressed here: Doesn't it go against the fundamentals of branding to be so agile and adapt to change so easily? What about the core things that make up a brand? For this reason, we need to think of the Agile Brand as being nuanced.

Earlier, we described "timelessness" as one of the key aspects of a great brand. It's important to make sure we don't confuse matters. While much of our discussion of the Agile Brand has centered on creating a continually evolving entity, we should also be clear that there are certain brand elements that should not be readily modified.

Thus, there are 2 components to an Agile Brand:

1. The things that don't change
2. The things that do change

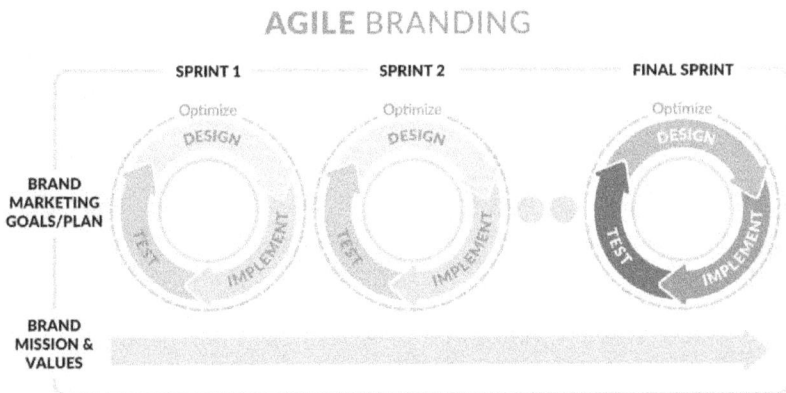

Figure 12.1, Agile Branding incorporates both Sprints as well as points of continuity

While your brand marketing and awareness efforts (top row) will inevitably change as you target new audiences or as your current audiences evolve, your brand mission and values (bottom row) should remain relatively constant over time.

GREG KIHLSTRÖM | 130

The things that don't (or seldom) change

Let's start with a deeper dive into the things that you should modify in
your Agile process. These include the elements that make your
company stand the test of time.

Mission

Despite your need to reach audiences in new ways, introduce products
and services, or even rebrand, your mission shouldn't be subject to as
frequent (or potentially as drastic) change.

Changing your mission should be done carefully, and it
requires greater justification than a change in customer preference or
marketing trends. After all, no matter what your logo looks like, what
your advertising says, or even who your target audience is, your
company should always stay true to its purpose and reason for
existence.

Values

Corporate values have been in the spotlight for many reasons lately. A
recent Nielsen poll[25] showed that 64% of consumers now consider
themselves belief-driven buyers, willing to spend more on a product if
the company that makes it shares their values[26].

John Wooden, a 12-year coach of the UCLA basketball team and
nicknamed the "Wizard of Westwood," said, "The true test of a man's
character is what he does when no one is watching." The same can be
said for a company's values.

Values are what drive the company independent of profit margins, shareholder value, and other business key performance indicators (KPIs).

The important thing about corporate values to keep in mind is that to be perceived as genuine and thus effective, they must remain constant. While different generations might have different ways of connecting with them, and companies may use ways of explaining them over time, what makes them ring true is their consistency and integrity.

No matter what trends and fads come and go, your values define who you are, the types of people you represent, and the type of future you envision, which is all great if your values are positive and inclusive and do not come at someone else's expense.

We can often assume that the word "values" is a positive thing, but many different people (and shareholders) find value in different things. While this isn't a book about politics or civil liberties, it's important to note that a company's values *should* be something positive that inspires its employees, its customers, and the communities it serves.

So, while an organization shouldn't change its values on a whim, it needs to be cognizant of the responsibility it has to people everywhere and the impact it has on the world.

As examples, here are three companies you are likely very familiar with that have stayed true to their mission despite changing times. You should note that these companies are *also* very agile and have made many evolutions over the years.

Patagonia

Patagonia, the outdoor clothing and gear company, has remained steadfast in its mission to "build the best product, cause no unnecessary harm, use business to inspire and implement solutions to the environmental crisis." Patagonia has consistently prioritized sustainability and environmental activism despite changing consumer behaviors and evolving market conditions. The company has taken bold steps, such as donating a significant portion of its profits to environmental causes and promoting the repair and reuse of its products. Even in the face of increasing competition and economic pressures, Patagonia's commitment to its environmental mission has resonated with eco-conscious consumers and solidified its position as a leader in sustainable business practices.

LEGO

LEGO has stayed true to its mission of "inspiring and developing the builders of tomorrow" by focusing on creativity, learning, and quality play experiences. Despite the rise of digital gaming and changing toy industry dynamics, LEGO has continually adapted its products and business model while maintaining its core mission. The company has successfully integrated digital experiences with physical play through products like LEGO Mindstorms and LEGO Dimensions, combining traditional building blocks with modern technology. LEGO's commitment to educational play has also led to partnerships with educational institutions and the development of programs that promote STEM learning, ensuring that its mission remains relevant in the digital age.

Ben & Jerry's

Ben & Jerry's, the iconic ice cream company, has always been driven by its mission to "create linked prosperity for everyone that's connected to our business: suppliers, employees, farmers, franchisees, customers, and neighbors alike." Despite the changing landscape of the food industry and shifting consumer preferences, Ben & Jerry's has remained committed to social justice, environmental sustainability, and fair trade practices. The company has continued using its platform to advocate for issues such as climate change, marriage equality, and racial justice while sourcing fair trade ingredients and supporting sustainable farming practices. This unwavering commitment to its mission has not only differentiated Ben & Jerry's in a crowded market but also built a loyal customer base that values its ethical stance.

The things that do change

Audience

There are two ways to look at how an audience changes over the life of a brand. First, you can look at audience shifts in terms of how an audience's preferences and behaviors change over time. This could be measured in everything from how they interact with your brand (e.g., a shift in mobile device usage or increased adoption of social media for customer service) to their buying behaviors or other preferences.

The second way you can look at audience shifts is by examining how different audiences may find your products and services useful over time. You may go to market assuming that your product solves a specific audience's problems but then find that you are instead being very successful with a completely different audience. What do you do? Depending on your strategic approach, you may decide to embrace this new audience, or you may shift your messaging and strategies to focus more on a different demographic.

Toyota's former Scion brand is a perfect example of this phenomenon. Toyota launched Scion as a low-cost, highly customizable brand of uniquely designed cars in 2003, which came out of its internally-designated "Project Genesis" started in 1999[27] to design and market cars to millennials. As of 2014, there were five models, from sporty to more utilitarian, and despite a heavy marketing focus on younger demographics, it failed to connect. Instead, Scion's models, and the boxy xB in particular, found a very different and older core audience, with Patrick George of Jalopnik[28] going so far as to say the models' success was "because Scions like this second-generation xB were affordable and easy for arthritic retirees to get in and out of." Despite a very targeted Millennial marketing effort, Gen Xers and Baby Boomers instead made up a significant portion of the brand's core customers, with some reporting the average age of a Scion customer was 49—far from the age of millennials.

Instead of embracing this and growing Scion in these segments, Toyota decided to shutter the brand on February 3, 2016, and fold its remaining successful car models into the parent brand[29]. Many say this was a missed opportunity. While the original audience didn't embrace Scions the way Toyota wished they would, the brand found a niche audience regardless.

Defining your audiences in terms of demographics can prove quite challenging and often relies on assumptions that because people are the same age or come from the same cultural background, they want the same things.

Because of this ambiguity, which will continue to increase with future generations, many marketers are moving away from targeting specific *demographic* audiences and moving toward targeting people at certain life moments. For example, two people of different ages and backgrounds who are both shopping for a new car loan for the first time actually have a lot in common.

Thinking of audiences in terms of commonality of purpose as opposed to the commonality of age, gender, culture, or generation affords marketers a very targeted way of looking at behavior and motivation.

This is also the type of approach that drives customer journey orchestration efforts, which focus on customer intent and their point in the overall customer or buyer's journey than it does on specific demographics.

Positioning

How companies position themselves relative to other competitors and even to their customers is something that *can* change over time.

Though a brand only wants to do so much of this, this quite often becomes an evolutionary change versus something that suddenly changes overnight.

Additionally, positioning may be something that evolves because of generational changes, such as the best way for a company that sells products for children to reach parents over time, as both their children and the parents themselves have different values than the generation that preceded them.

Strategies & tactics

Finally, the strategies used to tie your business goals to your audiences and the methods used to position your organization and its products and services will most definitely change over time. And for an Agile Brand, these things will be modified regularly.

Just think about the strategies that Coca-Cola uses now versus 45 years ago. Even 20 years ago, the Web was in its infancy and hardly the place to sell soft drinks. When the 1971 commercial featuring "I'd Like to Teach the World to Sing" aired, no one could have predicted that 45 years later, Coca-Cola would have a 55-person North American Social Centre[30] functioning as a real-time newsroom and social media marketing war room. While that is an obvious example because the medium (social media) simply didn't exist before this millennium, there are plenty of times that strategies and tactics have changed without changing the medium used.

Other Things to Consider

Let's end the chapter with a few additional things to keep in mind.

Unique selling proposition

Somewhere in between your values, which should never change, and your positioning, which may fluctuate with changing times and audience preferences, your unique selling proposition should not be lightly tossed aside, yet it may change relative to how some of the factors above ebb and flow over time.

Competition

Another constantly changing dynamic is the competitive landscape. Many companies have seen new and more disruptive competition from longstanding rivals as well as newer upstart companies with new and novel business models.

Think about how Amazon has disrupted the brick-and-mortar retail industry, how Uber has disrupted transportation, or how Stripe, Venmo, and other "fintech" or financial technology companies are disrupting more traditional banking, payment, and credit services.

Conclusion

The duality of the Agile Brand embodies the balance between structured principles and adaptive flexibility. By integrating formalized Agile practices with a broader mindset of agility, brands can create a dynamic and resilient approach to marketing and operations. This duality ensures that while there is a framework for efficiency and collaboration, there is also room for innovation, responsiveness, and continuous improvement. Embracing both aspects allows brands to stay competitive, foster stronger relationships with customers, and drive sustained growth.

Next Best Actions

1. **Conduct an Agile Audit:** Review your current Agile practices to assess their effectiveness and flexibility. Identify areas where rigid adherence to methodologies may be hindering adaptability. Ask: Are our Agile practices enhancing our ability to respond to change, or are they becoming too rigid?

2. **Create a Balance Strategy:** Develop a strategy that balances structured Agile methodologies with adaptive practices. Ensure that there is room for innovation and responsiveness within your Agile framework. Ask: How can we maintain the benefits of Agile principles while also fostering a culture of adaptability?

3. **Implement Feedback Loops:** Establish regular feedback loops from both internal teams and customers to refine and improve your Agile practices continuously. Use this feedback to make iterative adjustments and drive innovation. Ask: How can we

better incorporate feedback to enhance both our structured practices and our adaptability?

By focusing on these actions, organizations can effectively harness the duality of the Agile Brand, ensuring that they remain both efficient and flexible in a constantly evolving market. This balanced approach will drive long-term success and foster a culture of continuous improvement and innovation.

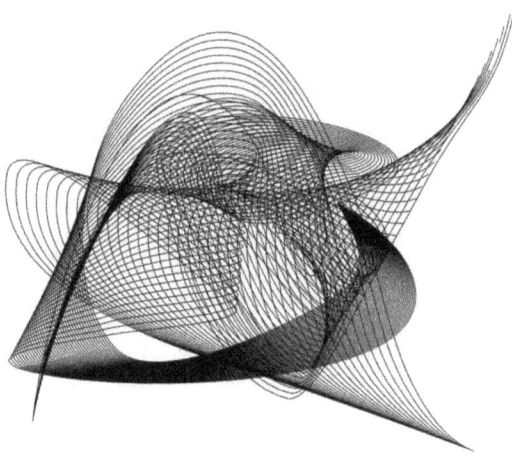

Chapter 16 |
Shared definition of
business value

TL;DR

Aligning all stakeholders on a shared definition of business value is crucial for an Agile Brand. When marketing, operations, technology, and finance all understand and pursue common principles, goals, and metrics, the organization can drive greater success. This alignment ensures coherence and collaboration across departments, allowing for more efficient decision-

making, resource allocation, and performance tracking.

In an Agile Brand, success hinges on a unified understanding of business value across all stakeholders. This means that marketing, operations, technology, and finance must not only recognize their unique contributions but also align on a common set of principles, goals, and metrics that define overall business value. When every department looks at value in similar ways yet respects each other's specialized performance metrics, the organization can achieve greater coherence, efficiency, and innovation.

The concept of business value can vary significantly between departments. Marketing may focus on customer engagement and brand perception, while operations might prioritize efficiency and process optimization. Technology departments often concentrate on innovation and system performance, whereas finance is driven by profitability and cost management. Despite these different perspectives, a shared definition of business value ensures that all efforts are aligned towards common objectives, fostering a collaborative environment where every team understands how their work contributes to the overall success of the organization.

This chapter explores the importance of a shared definition of business value, how to align principles, goals, and metrics across departments, and the benefits of fostering a unified approach. By embracing this holistic view, Agile Brands can navigate complexities more effectively, make informed decisions, and drive sustainable growth. Through case studies and practical strategies, we will illustrate how organizations can overcome common challenges and create a cohesive culture centered on shared value.

The Concept of Business Value

Business value is a multifaceted concept that can encompass a variety of elements depending on the perspective of different stakeholders within an organization. For an Agile Brand, business value includes not only financial performance but also customer satisfaction, operational efficiency, innovation, and employee engagement. Understanding business value as a holistic measure is crucial for aligning efforts across the organization and ensuring that every department works towards the same overarching goals.

In the context of an Agile Brand, business value can be broken down into several key components:

- **Customer Value:** The satisfaction and loyalty of customers based on their experiences with the brand's products and services.
- **Operational Value:** The efficiency and effectiveness of internal processes that contribute to delivering products and services.
- **Financial Value:** The profitability and cost-effectiveness of the organization, ensuring sustainable growth and return on investment.

- **Innovative Value:** The ability to continuously innovate and adapt to changing market conditions and customer needs.
- **Employee Value:** The engagement, satisfaction, and productivity of employees, which drive overall organizational performance.

The Impact of Differing Perspectives on Value Across Departments

Each department within an organization often has its own perspective on what constitutes business value, influenced by its specific roles and responsibilities. These differing perspectives can sometimes lead to misalignment and conflicts if not properly managed. For example:

- **Marketing** may focus on metrics such as brand awareness, customer acquisition, and engagement rates. Their primary concern is how customers perceive the brand and how effectively they can attract and retain them.
- **Sales** may focus on pipeline, likely and closed opportunities. Their primary concern is how the brand is able to overcome customer objections and that the product or service is of value to them.
- **Operations** might prioritize process efficiency, cost reduction, and quality control. Their goal is to ensure

that products and services are delivered in the most efficient and effective manner.

- **Technology** departments often emphasize innovation, system performance, and scalability. They look at value in terms of technological advancements and the ability to support current and future business needs.
- **Finance** is primarily concerned with profitability, cost management, and financial stability. They focus on the financial health of the organization and the return on investments.

While these perspectives are all valid, the challenge lies in ensuring that they do not operate in silos but rather contribute to a cohesive understanding of business value. This requires a concerted effort to align these diverse viewpoints around shared principles, goals, and metrics.

By defining business value in a way that resonates with all departments, an Agile Brand can foster a more integrated and collaborative approach. This alignment helps to break down silos, encourages cross-functional teamwork, and ensures that every action taken by individual departments supports the broader organizational objectives.

In the following sections, we will explore how to align principles, goals, and metrics across departments to create a unified understanding of business value. This alignment will not only enhance internal coherence but also drive greater overall success for the Agile Brand.

Aligning Principles Across Departments

To create a shared definition of business value, it's essential to establish core principles that resonate with all stakeholders across the organization. The seven principles of an Agile Brand we discussed earlier serve as the foundation for aligning efforts and ensuring that every department understands and supports the broader organizational goals.

Setting Common Goals

Setting common goals is critical for ensuring that all departments within an organization are aligned and working towards the same overarching objectives. Unified goals provide a clear direction and purpose, helping to coordinate efforts and resources across different functions. When everyone in the organization understands and commits to shared goals, it enhances coherence, reduces misalignment, and drives collective success.

Common goals serve several essential purposes:

1. **Direction:** They provide a clear path for all departments, ensuring that everyone knows where the organization is headed.
2. **Motivation:** Unified goals inspire and motivate employees by giving them a sense of purpose and shared mission.

3. **Measurement:** They establish benchmarks for performance, enabling the organization to track progress and make informed decisions.
4. **Collaboration:** Shared goals encourage cross-functional teamwork, breaking down silos and fostering a collaborative culture.

Developing and Communicating Shared Goals

To effectively set and communicate shared goals, organizations should follow a structured approach:

Engage Stakeholders:

- **What to Do:** Involve representatives from all departments in the goal-setting process to ensure that the goals reflect the diverse perspectives and needs of the organization.
- **How to Do It:** Conduct workshops, brainstorming sessions, and meetings to gather input and build consensus around the goals.

Align with Core Principles:

- **What to Do:** Ensure that the goals align with the organization's core principles, such as customer-centricity, innovation, and transparency.

- **How to Do It:** Review the established principles and verify that the goals support and reinforce these foundational values.

Set SMART Goals:

- **What to Do:** Develop goals that are Specific, Measurable, Achievable, Relevant, and Time-bound (SMART).
- **How to Do It:** Break down high-level objectives into detailed, actionable steps with clear metrics and deadlines.

Communicate Clearly and Consistently:

- **What to Do:** Share the goals widely and regularly within the organization to ensure that everyone understands and is committed to achieving them.
- **How to Do It:** Use multiple communication channels, such as emails, meetings, intranet updates, and visual dashboards, to keep the goals top of mind.

Foster Accountability:

- **What to Do:** Assign ownership of specific goals to departments or individuals to ensure accountability and follow-through.

- **How to Do It:** Set up regular check-ins and progress reviews to monitor achievement and address any obstacles.

By setting and communicating these shared goals, organizations can ensure that every department is aligned and working together towards the same objectives. This alignment enhances organizational coherence, fosters collaboration, and drives overall success.

In the next section, we will explore how to define unified metrics that reflect shared business value while accommodating the individualized performance metrics of each department.

Defining Unified Metrics

Unified metrics are essential for ensuring that all departments are aligned in their efforts to achieve the organization's goals. Key Performance Indicators (KPIs) serve as measurable values that reflect how effectively an organization is achieving its key business objectives. An Agile Brand can create a coherent framework for measuring success and driving continuous improvement by identifying KPIs that resonate across all departments.

To define KPIs that reflect shared business value, consider the following steps:

- **Align with Organizational Goals:** Ensure the KPIs are directly linked to the organization's overarching goals and core principles. This alignment ensures that every department's efforts contribute to the broader objectives.

- **Balance Quantitative and Qualitative Metrics:** Include both quantitative metrics (e.g., sales revenue, customer retention rates) and qualitative metrics (e.g., customer satisfaction, employee engagement) to provide a comprehensive view of performance.

- **Ensure Relevance Across Departments:** Select KPIs relevant to multiple departments and reflect the interconnected nature of their work. For example, customer satisfaction impacts marketing, operations, and customer service.

- **Focus on Actionable Metrics:** Choose KPIs that can be influenced by the actions of the departments. Metrics should be specific, measurable, and within the control of the teams responsible for achieving them.

- **Review and Adjust Regularly:** Periodically review and adjust the KPIs to ensure they remain relevant and aligned with evolving business objectives and market conditions.

Examples of Unified Metrics

Here are examples of how unified metrics can be established to reflect shared business value. Notice how, while each team has its own metrics, they all support the overall measure of success. This helps keep teams aligned and focused on the prize.

Customer Experience:

- **Overall Metric:** Customer Satisfaction (CSAT) Score or Net Promoter Score (NPS). Bonus points for those organizations that utilize Customer Lifetime Value (CLV) instead.
- **Marketing:** Customer engagement rates.
- **Operations:** Average order processing time.
- **Technology:** System response time for customer queries.
- **Finance:** Investment in customer service improvements.

Innovation and Growth:

- **Overall Metric:** Revenue from new products/services.
- **Marketing:** Number of new product launches.
- **Operations:** Efficiency improvements in new product production.
- **Technology:** Number of innovations implemented.
- **Finance:** Budget allocated to research and development.

Operational Efficiency:

- **Overall Metric:** Cost per unit produced.
- **Marketing:** Cost per lead acquisition.
- **Operations:** Production cycle time.

- **Technology:** IT infrastructure costs.
- **Finance:** Overall cost reduction percentage.

By defining and balancing unified metrics with individualized performance metrics, organizations can ensure that all departments are aligned in their efforts to achieve shared business value. This approach fosters a cohesive and collaborative culture, enhances transparency, and drives continuous improvement.

In the next section, we will explore strategies for fostering open communication and collaboration among departments to maintain alignment and transparency.

Communication and Collaboration

Effective communication and collaboration are crucial for maintaining alignment and ensuring that all departments are working towards common goals. By fostering a culture of openness and teamwork, organizations can break down silos, enhance transparency, and drive collective success. Here are key strategies to achieve this:

Examples of Successful Communication and Collaboration

While this may seem easier said than done, you can look to some other well-known examples for inspiration here. Let's explore a few companies that have embraced this shared definition of business value.

Spotify's Squad Model

Spotify uses a "squad" model where cross-functional teams, called squads, work on specific projects or features. Each squad operates autonomously but is aligned with the company's overall goals. This model fosters collaboration, innovation, and agility, enabling Spotify to quickly adapt to market changes and continuously improve its offerings.

Zappos' Open Communication Culture

Zappos emphasizes open communication and transparency. Regular all-hands meetings, accessible leadership, and a strong feedback culture are key aspects. This approach has helped Zappos maintain high levels of employee engagement, customer satisfaction, and operational efficiency.

By implementing these strategies and tools, organizations can foster a culture of open communication and collaboration, ensuring that all departments are aligned and working towards shared goals. This alignment enhances transparency, drives continuous improvement, and supports the overall success of the Agile Brand.

Common Challenges

Aligning all stakeholders around a shared definition of business value is a complex task that often encounters several challenges. Understanding these challenges and proactively addressing them can help ensure successful alignment. Here are some common challenges and strategies to overcome them:

Siloed Departments

- **Challenge:** Departments often operate in silos, focusing on their own goals and metrics without considering the broader organizational objectives.
- **Solution:** Implement cross-functional teams and regular inter-departmental meetings to foster collaboration. Encourage leaders to promote a culture of openness and teamwork across the organization.

Resistance to Change

- **Challenge:** Employees and departments may resist changes to established processes and practices, fearing disruption or uncertainty.
- **Solution:** Communicate the benefits of alignment clearly and provide training to help employees adapt. Involve stakeholders in the change process to increase buy-in and reduce resistance.

Lack of Clear Communication

- **Challenge:** Misalignment often stems from poor communication, where goals and metrics are not clearly articulated or understood across departments.
- **Solution:** Develop a comprehensive communication plan that includes regular updates, clear documentation, and accessible

channels for feedback and questions. Use visual aids like dashboards to make information easily understandable.

Different Metrics and Priorities

- **Challenge:** Departments may have conflicting metrics and priorities that hinder alignment.
- **Solution:** Establish unified KPIs that reflect shared business value while respecting individual departmental metrics. Ensure that these KPIs are aligned with the overall business objectives and communicated effectively.

Limited Resources

- **Challenge:** Aligning stakeholders requires time, effort, and resources, which may be limited.
- **Solution:** Prioritize alignment activities that have the highest impact on business value. Allocate resources strategically and seek efficiencies through collaboration and technology.

Approaches Overcome Resistance and Maintain Alignment

Of course, not everyone is going to be "on board" on day one. It takes time, effort, and a good plan to overcome initial resistance, as well as to ensure that teams stay aligned on this common, shared purpose. Here are a few ways to do this.

- **Leadership Commitment:** Ensure that leaders at all levels are committed to the alignment process and model the desired behaviors. Leadership should actively participate in alignment activities and demonstrate their importance through actions and communication.

- **Ongoing Training and Development:** Provide continuous training and development opportunities to help employees understand and embrace the shared definition of business value. Offer workshops, seminars, and online courses to build skills and knowledge.

- **Feedback and Iteration:** Establish mechanisms for regular feedback and iteration. Encourage employees to share their experiences and suggestions for improvement. Use this feedback to refine processes and maintain alignment over time.

- **Celebrating Success:** Recognize and celebrate achievements that result from aligned efforts. Highlight case studies and success stories to demonstrate the positive impact of alignment on business outcomes.

- **Adaptability and Flexibility:** Maintain flexibility in your approach to alignment. Be willing to adapt strategies and processes based on feedback and changing circumstances. This adaptability will help sustain alignment and ensure it remains relevant and effective.

Conclusion

A shared definition of business value is essential for the success of an Agile Brand. When all stakeholders, from marketing to finance, align on common principles, goals, and metrics, the organization can operate more efficiently and effectively. This alignment fosters a cohesive culture, enhances decision-making, and drives continuous improvement. By maintaining a unified understanding of what constitutes value, Agile Brands can navigate the complexities of the market with greater agility and resilience.

Next Best Actions

1. **Conduct a Stakeholder Alignment Workshop:**
   - Organize workshops with representatives from all departments to discuss and define shared principles, goals, and metrics. Foster open dialogue to ensure everyone's perspectives are considered. Ask: How can we create a unified understanding of value that aligns with our organizational goals?

2. **Develop and Communicate a Unified Value Framework:**
   - Create a comprehensive framework that outlines the shared definition of business value, including common principles, goals, and KPIs. Communicate this framework across the organization through training sessions, internal communications, and ongoing meetings. Ask: How can we ensure all stakeholders

understand and embrace our shared definition of value?

3. **Implement Regular Cross-Departmental Reviews:**
    o Establish regular meetings where representatives from different departments review performance metrics, share insights, and discuss alignment with shared goals. Use these reviews to identify areas of misalignment and take corrective actions. Ask: How can we maintain continuous alignment and address any emerging discrepancies in our understanding of value?

By following these actions, organizations can embed a shared definition of business value into their operations, ensuring cohesive and collaborative efforts towards common objectives. This alignment will drive greater success, innovation, and agility in the face of evolving market challenges.

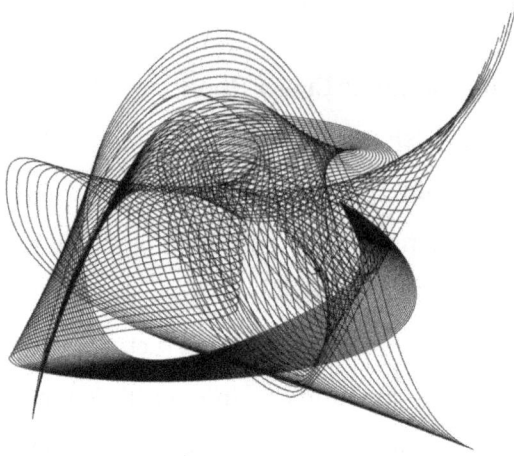

Chapter 17 | Culture of Experimentation

TL;DR

Creating a culture of experimentation empowers organizations to embrace agility, foster innovation, and continuously improve. By encouraging experimentation, companies can better adopt the seven principles of an Agile Brand, driving sustained success and adaptability in a rapidly changing market.

A culture of experimentation is an organizational environment that encourages continuous testing, learning, and innovation. In this culture, employees are empowered to propose, design, and execute experiments aimed at improving processes, products, services, and customer experiences. Experimentation is not considered a risky venture but a necessary practice for growth and adaptation.

Experimentation encourages creative thinking and the exploration of new ideas. By testing hypotheses and learning from outcomes, organizations can discover innovative solutions and improvements. A culture of experimentation supports agility by allowing organizations to adapt to changes and uncertainties quickly. By regularly testing and iterating, companies can respond more effectively to market shifts and emerging trends.

A culture of experimentation also has the following benefits:

1. **Drives Continuous Improvement:**
   - Continuous experimentation leads to ongoing enhancements in processes and products. This iterative approach ensures that the organization is always moving forward, refining its offerings, and improving efficiency.

2. **Promotes Employee Engagement:**
   - Employees who are encouraged to experiment and innovate feel more invested in their work and the company's success. This engagement leads to higher job satisfaction and retention.

3. **Reduces Fear of Failure:**
   - In a culture that embraces experimentation, failure is viewed as a valuable learning opportunity. This

perspective reduces the fear of taking risks and encourages employees to try new approaches without fear of negative consequences.

The Relationship Between Experimentation and Innovation

Experimentation is the engine that drives innovation. By systematically testing new ideas and approaches, organizations can identify what works and what doesn't, leading to breakthrough innovations. This process involves:

- **Hypothesis Formation:** Developing new ideas based on observations and insights.
- **Experiment Design:** Creating structured tests to validate or invalidate these hypotheses.
- **Data Collection:** Gathering and analyzing data to draw meaningful conclusions.
- **Iteration:** Refining and improving ideas based on experimental outcomes.

How it supports the 7 Agile Brand principles

A culture of experimentation aligns perfectly with the principles of an Agile Brand. It provides the flexibility and responsiveness needed to adapt quickly to changes, fosters continuous improvement, and encourages innovation. By embedding experimentation into the organizational culture, Agile Brands can:

- **Agility by Design:** Build systems and processes that are inherently flexible and adaptable.
- **Continuous Improvement:** Use experimentation to identify incremental and transformative changes.
- **Operationalizing Adaptivity:** Make experimentation a core part of daily operations.
- **Guided by Values:** Ensure experiments align with the organization's values and ethical standards.
- **Building Relationships:** Enhance customer and employee relationships through meaningful and innovative interactions.
- **Focusing on the Conversation:** Engage stakeholders in the experimentation process to gather insights and feedback.
- **Always Learning and Growing:** Use experimentation as a tool for continuous learning and development.

By understanding and embracing a culture of experimentation, organizations can create a dynamic environment that supports innovation, agility, and continuous improvement. This foundation enables companies to thrive in a rapidly changing market and maintain their competitive edge.

Aligning Experimentation with Agile Brand Principles

Agility by design means structuring your organization to be flexible and responsive. In the context of experimentation, this involves creating processes that allow for quick iteration and adaptation based on experimental outcomes.

How it supports the 7 Agile Brand principles

- **Flexible Processes:** Designing experiments that can be easily adjusted based on initial results helps maintain momentum and ensures continuous learning.
- **Rapid Prototyping:** Encouraging rapid prototyping and testing of ideas allows organizations to pivot quickly and effectively.

Implementation example

A software company implements A/B testing to continuously improve user experience. By rapidly testing different interface designs and features, they can quickly adapt to user preferences and market trends.

Continuously Improving: Using Experimentation to Drive Incremental and Transformative Changes

Continuous improvement involves making regular, incremental enhancements while also being open to transformative changes when necessary. Experimentation provides a systematic approach to identifying and implementing these improvements.

How it supports the 7 Agile Brand principles

- **Incremental Enhancements:** Regular experimentation helps identify small but impactful changes that improve processes and products over time.
- **Transformative Innovation:** Larger, more ambitious experiments can lead to breakthrough innovations that significantly alter the organization's trajectory.

Implementation example

A manufacturing firm regularly experiments with process improvements on the production line, leading to steady increases in efficiency and quality. Occasionally, they undertake larger experiments with new manufacturing technologies, resulting in significant leaps in capability.

Operationalizing Adaptivity: Making Experimentation a Core Part of Business Operations

Operationalizing adaptivity means embedding experimentation into the everyday operations of the organization, ensuring that it becomes a routine part of how the business functions.

How it supports the 7 Agile Brand principles

- **Routine Testing:** Making experimentation a regular part of operations ensures that the organization remains adaptive and responsive.
- **Standardized Procedures:** Establishing standardized procedures for experimentation integrates this practice into the organizational workflow.

Implementation example

A retail company uses continuous experimentation in its e-commerce platform to test different pricing strategies, website layouts, and promotional offers. This ongoing testing helps them stay competitive and responsive to consumer behavior.

Guided by Values: Ensuring Experiments Align with Organizational Values and Ethics

Being guided by values involves ensuring that all experiments align with the core principles and ethical standards of the organization. This alignment guarantees that innovation does not come at the expense of integrity.

- **Ethical Considerations:** Ensuring that experiments adhere to ethical guidelines builds trust and maintains the organization's reputation.
- **Value-Driven Goals:** Designing experiments that reflect the organization's values ensures that outcomes contribute positively to the brand's mission.

Implementation example

A tech company committed to data privacy conducts experiments on user data only with explicit consent and transparency, maintaining trust and upholding its ethical standards.

Building Relationships: Leveraging Experimentation to Enhance Customer and Employee Relationships

Building relationships through experimentation involves engaging customers and employees in the process and using their feedback to guide improvements.

How it supports the 7 Agile Brand principles

- **Customer Engagement:** Involving customers in experiments, such as beta testing new features, strengthens the relationship and ensures that products meet their needs.

- **Employee Involvement:** Encouraging employees to propose and participate in experiments fosters a sense of ownership and engagement.

Implementation example

A consumer goods company invites loyal customers to test new product variations and provide feedback, which helps refine the products and strengthens customer loyalty.

Focusing on the Conversation: Engaging Stakeholders in the Experimentation Process

Focusing on the conversation means actively involving stakeholders in the experimentation process and encouraging open dialogue and feedback.

How it supports the 7 Agile Brand principles

- **Stakeholder Feedback:** Regularly seeking input from stakeholders ensures that experiments are relevant and aligned with their needs and expectations.
- **Transparent Communication:** Keeping stakeholders informed about the purpose, process, and outcomes of experiments builds trust and support.

Implementation example

A financial services company involves clients in the development of new financial products by seeking their input through surveys and focus groups, ensuring that the final offerings align with client needs.

Always Learning and Growing: Using Experimentation as a Tool for Continuous Learning

Always learning and growing involves using the insights gained from experiments to foster continuous improvement and development across the organization.

How it supports the 7 Agile Brand principles

- **Learning from Outcomes:** Analyzing experimental results to understand what works and what doesn't drives ongoing improvement.
- **Knowledge Sharing:** Sharing the lessons learned from experiments across the organization promotes collective growth and development.

Implementation example

A healthcare organization conducts regular experiments on patient care processes, analyzing the outcomes to identify best practices and disseminating these insights throughout the network to improve overall care quality.

By aligning experimentation with the principles of an Agile Brand, organizations can foster a culture of continuous improvement, innovation, and adaptability. This approach ensures that the organization remains responsive to changes and committed to delivering value to customers, employees, and other stakeholders.

Key Elements of a Culture of Experimentation within an Organization

So this sounds pretty great, right? How does an organization adopt a culture of experimentation, and what are the elements needed to keep the pace going over time? Let's explore several of these.

Encouraging Curiosity and Creativity

Encouraging curiosity and creativity involves fostering an environment where employees feel empowered to ask questions, explore new ideas, and think outside the box. This culture supports innovative thinking and allows for the generation of unique solutions and approaches.

How it supports the 7 Agile Brand principles

- **Innovation Drive:** By nurturing curiosity, organizations can unlock new ideas and opportunities that drive innovation.

- **Employee Engagement:** Encouraging creativity makes employees feel valued and motivated, leading to higher engagement and job satisfaction.

Implementation example

A marketing firm holds regular brainstorming sessions where team members can present and discuss creative ideas without judgment. These sessions often lead to innovative campaigns and strategies.

Providing Resources and Tools

Providing resources and tools means equipping employees with the necessary technology, time, and support to conduct experiments. This investment ensures that experimentation is feasible and effective.

How it supports the 7 Agile Brand principles

- **Effective Experimentation:** Adequate resources enable employees to design and execute meaningful experiments that yield valuable insights.
- **Empowerment:** Providing tools and resources demonstrates trust in employees' abilities to innovate and improve processes.

Implementation example

A software company provides access to advanced analytics tools and dedicated time for developers to experiment with new coding techniques, leading to improved software performance and innovation.

Promoting Psychological Safety

Promoting psychological safety involves creating a work environment where employees feel safe to take risks, make mistakes, and learn from failures without fear of negative repercussions.

How it supports the 7 Agile Brand principles

- **Encouraging Risk-Taking:** When employees feel safe to experiment, they are more likely to take calculated risks that can lead to significant innovations.
- **Learning from Failure:** A safe environment encourages learning from mistakes, which is essential for continuous improvement.

Implementation example

A financial services firm has a policy that no employee will be penalized for failed experiments as long as they document their learning. This approach has led to numerous successful innovations and improvements.

Establishing Clear Processes

Establishing clear processes involves creating structured methodologies for designing, executing, and evaluating experiments. This approach ensures consistency and reliability in the experimentation process.

How it supports the 7 Agile Brand principles

- **Consistency:** Clear processes provide a standard framework for experimentation, ensuring that all experiments are conducted systematically and can be replicated.
- **Efficiency:** Structured processes streamline the experimentation workflow, making it easier to manage and evaluate multiple experiments.

Implementation example

A healthcare organization develops a standardized process for testing new patient care protocols, including steps for planning, execution, data collection, and analysis. This process ensures that experiments are thorough and results are reliable.

Measuring and Sharing Results

Measuring and sharing results involves tracking the outcomes of experiments and disseminating the findings across the organization.

This practice ensures that valuable insights are not siloed but contribute to collective knowledge and improvement.

How it supports the 7 Agile Brand principles

- **Data-Driven Decisions:** Measuring results provides objective data that can guide decision-making and validate hypotheses.
- **Knowledge Sharing:** Sharing insights from experiments fosters a culture of learning and helps other teams apply successful strategies.

Implementation example

An e-commerce company uses a centralized dashboard to display the results of various marketing experiments, allowing teams across the organization to see what strategies are working and apply these insights to their own projects.

By embedding these key elements into the organizational culture, companies can create a robust framework for experimentation that supports continuous improvement and innovation. This culture not only drives the adoption of Agile Brand principles but also ensures that the organization remains dynamic, resilient, and competitive in a rapidly changing market.

Strategies That Leaders Should Adopt to Foster Experimentation

To grow and foster this culture of experimentation, commitment from leadership and other stakeholders is required. Let's look at what is needed.

Leadership Support

I'll start with the obvious, but because it is so important and yet still often missing so often, I think it bears repeating here. Leadership support is critical to building this, and it involves senior leaders actively championing experimentation within the organization. Leaders should model experimental behavior, encourage risk-taking, and provide the necessary resources and support for experimentation.

How it supports the 7 Agile Brand principles

- **Encouragement from the Top:** When leaders advocate for and participate in experimentation, it sets a positive example and encourages all employees to embrace this mindset.
- **Resource Allocation:** Leadership support ensures adequate resources are dedicated to experimentation, reinforcing its importance within the organization.

The CEO of a tech startup regularly participates in hackathons and innovation challenges alongside employees, demonstrating a commitment to experimentation and innovation.

Cross-Functional Collaboration

Cross-functional collaboration involves bringing together diverse teams from different departments to work on experiments. This approach leverages a wide range of skills, perspectives, and expertise to drive innovative solutions.

How it supports the 7 Agile Brand principles

- **Diverse Perspectives:** Collaboration across departments fosters creativity and leads to more comprehensive and innovative solutions.
- **Shared Learning:** Cross-functional teams promote knowledge sharing and help break down silos within the organization.

Implementation example

A consumer electronics company forms cross-functional teams comprising members from R&D, marketing, and customer service to develop and test new product concepts. This collaboration results in products that better meet customer needs and preferences.

Training and Development

Training and development involve providing employees with the skills and knowledge they need to design and execute experiments effectively. This includes training on experimental design, data analysis, and innovation methodologies.

How it supports the 7 Agile Brand principles

- **Skill Enhancement:** Training equips employees with the necessary tools to conduct meaningful experiments, driving better outcomes.
- **Continuous Improvement:** Ongoing development opportunities ensure employees stay current with the latest techniques and best practices.

Implementation example

A pharmaceutical company offers regular workshops on experimental design and statistical analysis, ensuring that researchers are well-equipped to conduct robust and reliable experiments.

Incentivizing Innovation

Incentivizing innovation involves recognizing and rewarding employees who contribute to a culture of experimentation. This can include financial rewards, recognition programs, and career advancement opportunities.

- **Motivation:** Incentives encourage employees to engage in experimentation and innovation, driving higher levels of participation and effort.
- **Recognition:** Recognizing innovative efforts reinforces the value of experimentation and motivates others to contribute.

Implementation example

A software development firm awards an "Innovator of the Month" title to employees who propose and successfully implement impactful experiments. This recognition is accompanied by a bonus and public acknowledgment at company meetings.

Embedding Experimentation in Daily Operations

Embedding experimentation in daily operations involves making experimentation a regular part of the workflow. This means integrating experimental thinking into everyday tasks and decision-making processes.

How it supports the 7 Agile Brand principles

- **Normalization:** Regular experimentation becomes a standard practice, fostering a culture where innovation and continuous improvement are routine.

- **Efficiency:** Embedding experimentation streamlines the process, making it an integral part of how the organization operates.

Implementation example

A retail chain incorporates small-scale experiments into its daily operations, such as testing different store layouts or promotional strategies in select locations. The insights gained are used to optimize practices across all stores.

By implementing these strategies, organizations can foster a culture of experimentation that supports the principles of an Agile Brand. Leadership support, cross-functional collaboration, training and development, incentivizing innovation, and embedding experimentation into daily operations are all critical components that drive continuous improvement, innovation, and adaptability. This culture ensures that the organization remains dynamic and responsive to changing market conditions, ultimately leading to sustained success.

Conclusion

Creating a culture of experimentation is fundamental to becoming an Agile Brand. This culture empowers organizations to continuously innovate, adapt, and improve, ensuring they remain competitive and responsive in a rapidly changing market. By embracing experimentation, companies can better align with the seven principles of an Agile Brand: Agility by Design, Continuously Improving,

Operationalizing Adaptivity, Guided by Values, Building Relationships, Focusing on the Conversation, and Always Learning and Growing.

A culture of experimentation fosters an environment where curiosity and creativity thrive, supported by the necessary resources, psychological safety, and structured processes. It encourages collaboration across departments, leveraging diverse perspectives to drive innovative solutions. Leadership support, ongoing training, and appropriate incentives further reinforce this culture, embedding experimentation into the organization's DNA.

By aligning experimentation with core principles, setting clear processes, and sharing results, organizations can create a robust framework that supports continuous improvement and innovation. This approach not only drives the adoption of Agile Brand principles but also ensures that the organization remains dynamic, resilient, and successful in the long term.

Next Best Actions

1. **Initiate an Experimentation Program:**
   - Launch a program that encourages employees to propose and conduct experiments related to their work. Provide guidelines and support for designing and executing these experiments. Ask: How can we create a structured yet flexible program that fosters experimentation?
2. **Promote a Learning Mindset:**

○ Develop training sessions and workshops emphasizing the value of learning from successful and failed experiments. Encourage a growth mindset throughout the organization. Ask: What resources and training can we provide to cultivate a learning-focused culture?

3. **Create a Feedback and Sharing Platform:**

○ Establish a platform where employees can share the results and insights from their experiments. Use this platform to facilitate cross-departmental learning and collaboration. Ask: How can we ensure that the knowledge gained from experiments is effectively shared and utilized across the organization?

By implementing these actions, organizations can foster a culture of experimentation that supports the principles of an Agile Brand, driving continuous improvement, innovation, and adaptability.

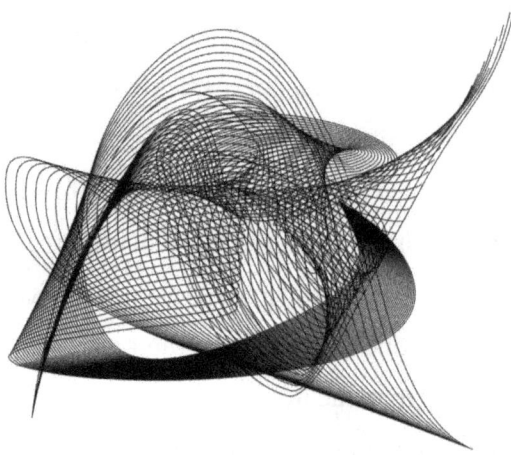

Chapter 18 | Feedback loop

TL;DR

Creating a robust feedback loop is critical to the long-term success of an Agile Brand. A comprehensive feedback loop should encompass performance metrics across marketing, customer experience, sales, and product performance, as well as data on team collaboration, process efficiency, customer satisfaction, and overall business value. This holistic approach ensures continuous improvement, fosters innovation, and enhances both customer and employee lifetime value.

Feedback is the lifeblood of an Agile Brand, providing the insights needed to drive continuous improvement and innovation. By creating a robust feedback loop, organizations can ensure they remain responsive to market changes, customer needs, and internal dynamics. This chapter explores the importance of a comprehensive feedback loop, encompassing performance metrics across various areas and providing a holistic view of organizational health and success.

Understanding the Feedback Loop

A feedback loop is a systematic process for collecting, analyzing, and acting on data to improve performance. It involves gathering information from various sources, interpreting the data to gain insights, and implementing changes based on these insights. For an Agile rand, a feedback loop is crucial for maintaining agility, fostering innovation, and achieving long-term success.

In the following sections, we will delve into the key areas of performance measurement, the importance of internal performance metrics, and how to integrate these metrics to create a holistic view of business value. We will also explore best practices for implementing an effective feedback loop and present case studies of successful Agile Brands that have leveraged feedback to drive continuous improvement.

Understanding the Feedback Loop

A feedback loop is a cyclical process of collecting data, analyzing it, and implementing changes based on the insights gained. This process helps organizations to continuously refine their strategies, improve their operations, and enhance overall performance. In an Agile Brand, a feedback loop is essential for maintaining flexibility, responsiveness, and innovation.

Components of a Feedback Loop:

1. **Data Collection:**
   - Gathering quantitative and qualitative data from various sources, including customer feedback, employee input, performance metrics, and market trends.
2. **Data Analysis:**
   - Interpreting the collected data to identify patterns, trends, and insights. This step involves using analytical tools and methodologies to make sense of the information.
3. **Action Planning:**
   - Developing strategies and plans based on the insights gained from the data analysis. This involves setting clear goals and identifying specific actions to address areas for improvement.
4. **Implementation:**

o   Executing the action plans and making the necessary changes to processes, products, or strategies. This step requires effective project management and collaboration across departments.

5. **Evaluation:**

     o   Monitoring the impact of the implemented changes and assessing their effectiveness. This involves tracking key performance indicators (KPIs) and gathering additional feedback to determine if the desired outcomes have been achieved.

6. **Iteration:**

     o   Repeating the process to ensure continuous improvement. The feedback loop is ongoing, with each cycle building on the previous one to drive incremental and transformative changes.

The Role of Feedback in Continuous Improvement and Agility

Continuous Improvement: Feedback is the foundation of continuous improvement, providing the information needed to identify strengths and weaknesses in an organization's operations, products, and strategies. By regularly collecting and analyzing feedback, organizations can make informed decisions that lead to better performance and higher quality.

- **Example:** A software company collects user feedback after each product release to identify bugs and areas for enhancement. This feedback is used to make iterative improvements in subsequent releases, ensuring that the product continuously evolves to meet user needs.

Agility: A robust feedback loop enhances an organization's agility by enabling it to respond quickly to changes in the market, customer preferences, and internal dynamics. By staying attuned to feedback, Agile Brands can pivot their strategies and operations as needed to stay competitive and relevant.

- **Example:** A retail chain uses real-time sales data and customer feedback to adjust its inventory and marketing strategies on the fly. This responsiveness allows the company to capitalize on emerging trends and customer demands, maintaining a competitive edge.

Driving Innovation: Feedback loops are also crucial for fostering innovation. Organizations can identify successful innovations and scale them effectively by experimenting with new ideas and gathering feedback on their performance.

- **Example:** A tech startup conducts A/B testing on new features and gathers user feedback to determine which features resonate most with users. The insights gained from these experiments guide the company's product development roadmap, ensuring that innovations align with user needs and preferences.

Building Trust and Engagement: Regularly seeking and acting on feedback demonstrates that an organization values the input of its customers and employees. This approach builds trust and engagement, fostering a positive relationship between the organization and its stakeholders.

- **Example:** A financial services company implements a feedback system where clients can rate their experience and suggest improvements. By acting on this feedback and communicating the changes made, the company builds stronger client relationships and enhances customer satisfaction.

A feedback loop is a vital mechanism for driving continuous improvement, agility, and innovation within an Agile Brand. By systematically collecting, analyzing, and acting on feedback, organizations can remain responsive to changes and committed to delivering value to their customers and employees. In the following sections, we will explore key areas of performance measurement and how to integrate these metrics to create a comprehensive feedback loop.

Holistic Business Value Metrics

To ensure long-term success and continuous improvement, an Agile Brand must integrate both external and internal performance metrics to create a comprehensive feedback loop. This holistic approach provides a complete view of business value by encompassing customer lifetime value (CLV), employee lifetime value (ELV), and overall business value. By combining these metrics, organizations can make informed decisions that drive sustained growth and innovation.

Understanding the Long-Term Value of Customers

Customer Lifetime Value (CLV) is a measure of the total revenue a company can expect to generate from a customer over the entire duration of their relationship. Understanding CLV helps organizations focus on strategies that enhance customer retention, satisfaction, and profitability.

Key Metrics

- **Average Purchase Value:** The average amount spent by a customer per purchase.
- **Purchase Frequency Rate:** The average number of purchases made by a customer over a specific period.
- **Customer Lifespan:** The average duration a customer remains active with the company.

Implementation example

An e-commerce company calculates CLV by analyzing customer purchase data and identifying patterns in buying behavior. This information helps the company develop targeted marketing campaigns and personalized offers to increase customer retention and maximize lifetime value.

Assessing the Long-Term Contributions of Employees

Employee Lifetime Value (ELV) measures the total value an employee brings to the organization over the course of their employment. This includes their contributions to productivity, innovation, and company culture. Understanding ELV helps organizations invest in employee development and retention strategies that enhance overall performance.

Key Metrics

- **Employee Productivity:** The output and efficiency of an employee's work.
- **Innovation Contributions:** The number and impact of innovative ideas and projects initiated by the employee.
- **Retention Rate:** The length of time an employee stays with the company.

Implementation example

A technology firm tracks ELV by monitoring employee performance metrics and innovation contributions. By offering professional development opportunities and recognizing top performers, the company boosts employee satisfaction and retention, ultimately increasing ELV.

Integrating Financial Performance, Market Share, and Growth Metrics

Overall business value encompasses a broad range of metrics that provide a comprehensive view of the company's financial health, market position, and growth potential. Integrating these metrics helps organizations make strategic decisions that drive long-term success.

Key Metrics

- **Revenue Growth:** The rate at which the company's revenue is increasing over time.
- **Market Share:** The company's share of the total market in which it operates.
- **Profit Margins:** The difference between revenue and costs, indicating the company's profitability.

Implementation example

A retail chain integrates financial performance metrics with market share and growth data to create a holistic view of business value. This comprehensive analysis informs strategic decisions on expansion, marketing investments, and operational improvements.

Integrating Metrics for a Holistic View

Integrating CLV, ELV, and overall business value metrics provides a complete picture of organizational performance. This holistic view allows companies to balance short-term goals with long-term strategies, ensuring sustained growth and success.

By combining these metrics, organizations can identify key drivers of value and allocate resources more effectively. This integrated approach supports strategic decision-making, enabling companies to prioritize initiatives that enhance both customer and employee satisfaction.

A holistic feedback loop facilitates continuous improvement by providing comprehensive insights into all aspects of the business. Regularly reviewing and updating these metrics ensures that the organization remains agile and responsive to changes in the market and internal dynamics.

Implementation example

A global consumer goods company integrates CLV, ELV, and overall business value metrics into a centralized dashboard. This dashboard is used by senior leadership to track performance, identify trends, and make informed decisions that drive long-term growth and profitability.

By focusing on holistic business value metrics, organizations can create a comprehensive feedback loop that drives continuous improvement and supports long-term success. In the next section, we will discuss best practices for implementing an effective feedback loop and present case studies of successful Agile Brands that have leveraged feedback to achieve sustained growth.

Implementing a Feedback Loop

Effective leadership is crucial for successfully implementing a comprehensive feedback loop within an organization. Leaders set the tone for a culture of continuous improvement and play a pivotal role in ensuring that feedback mechanisms are established, maintained, and utilized effectively. Here are the steps leaders can take to implement a robust feedback loop.

1. Establish Clear Objectives and Goals

Leaders must define the purpose and objectives of the feedback loop. Clear goals ensure that all stakeholders understand the importance of feedback and how it will be used to drive improvements.

Steps to Take:

- **Define Purpose:** Articulate why the feedback loop is being implemented and what the organization hopes to achieve.
- **Set Clear Goals:** Establish specific, measurable, achievable, relevant, and time-bound (SMART) goals for the feedback loop.
- **Communicate Objectives:** Ensure that all employees understand the goals and how their feedback will contribute to achieving them.

Implementation example

The CEO of a manufacturing company sets clear objectives for a new feedback loop aimed at improving production efficiency and product quality. These objectives are communicated through company-wide meetings and written communications.

2. Foster a Culture of Openness and Trust

A culture of openness and trust is essential for encouraging employees to provide honest and constructive feedback. Leaders must create an environment where feedback is valued and acted upon.

Steps to Take:

- **Lead by Example:** Demonstrate openness to feedback by actively seeking input from employees and acting on it.
- **Encourage Transparency:** Promote transparent communication and ensure that feedback processes are clear and accessible.
- **Build Trust:** Show employees that their feedback is valued and that there are no negative repercussions for providing honest input.

Implementation example

A tech startup's leadership team holds regular town hall meetings where employees can voice their opinions and suggestions. Leaders actively listen, address concerns, and provide updates on how feedback is being used to make improvements.

3. Implement Feedback Mechanisms

Leaders must establish formal mechanisms for collecting, analyzing, and acting on feedback. These mechanisms ensure that feedback is systematically gathered and used to drive improvements.

Steps to Take:

- **Select Appropriate Tools:** Choose tools and platforms that facilitate easy collection and analysis of feedback (e.g., surveys, suggestion boxes, feedback apps).
- **Standardize Processes:** Develop standardized processes for collecting and analyzing feedback to ensure consistency and reliability.
- **Provide Training:** Offer training to employees on how to use feedback tools and the importance of providing constructive feedback.

Implementation example

A retail chain implements an employee feedback app that allows staff to submit suggestions and concerns anonymously. Leaders provide training sessions on how to use the app and emphasize the importance of feedback for continuous improvement.

4. Analyze and Interpret Feedback

Collecting feedback is only the first step. Leaders must ensure that the feedback is thoroughly analyzed and interpreted to identify trends, issues, and opportunities for improvement.

Steps to Take:

- **Assign Responsibility:** Designate teams or individuals responsible for analyzing feedback data.
- **Use Analytical Tools:** Employ data analysis tools and methodologies to make sense of the feedback and extract actionable insights.
- **Report Findings:** Regularly report findings to relevant stakeholders and incorporate insights into strategic planning.

Implementation example

The HR department of a healthcare organization is tasked with analyzing employee feedback surveys. They use statistical analysis software to identify trends and present their findings to the leadership team for action.

5. Act on Feedback and Close the Loop

For a feedback loop to be effective, it is crucial to act on the feedback received and communicate the actions taken back to the employees.

This "closing the loop" demonstrates that feedback is valued and leads to tangible improvements.

Steps to Take:

- **Develop Action Plans:** Create specific action plans based on feedback insights, outlining steps, timelines, and responsible parties.
- **Implement Changes:** Execute the action plans and make the necessary changes to processes, products, or strategies.
- **Communicate Actions:** Inform employees about the changes made in response to their feedback and the expected impact.

Implementation example

A financial services firm acts on client feedback by enhancing its online banking platform. The company then communicates the updates through newsletters and client meetings, highlighting how the feedback led to improvements.

6. Monitor and Iterate

Leaders should continuously monitor the effectiveness of the feedback loop and make adjustments as needed. This iterative process ensures that the feedback loop remains relevant and effective over time.

Steps to Take:

- **Track Progress:** Regularly review the progress of action plans and the impact of changes implemented.
- **Gather Ongoing Feedback:** Continuously collect feedback to assess the effectiveness of the improvements and identify new areas for enhancement.
- **Adjust Strategies:** Refine feedback mechanisms and strategies based on ongoing feedback and changing organizational needs.

Implementation example

A software development company tracks the success of product updates based on user feedback and usage data. They continuously gather feedback through beta testing and make iterative improvements to ensure the product meets user expectations.

By taking these steps, leaders can successfully implement a feedback loop that drives continuous improvement and long-term success. This proactive approach ensures that the organization remains agile, responsive, and committed to delivering value to both customers and employees.

Common Roadblocks and Strategies to Overcome Them

Implementing a feedback loop is just the beginning; sustaining it in the long term can be challenging. Common roadblocks can hinder the effectiveness and continuity of feedback mechanisms. Identifying and developing strategies to overcome these challenges is essential for maintaining a robust feedback loop that drives continuous improvement and long-term success.

Feedback Fatigue

Feedback fatigue occurs when employees or customers become overwhelmed by the frequency or complexity of feedback requests, leading to decreased participation and engagement.

Strategies to Overcome It:

- **Optimize Frequency:** Balance the frequency of feedback requests to avoid overwhelming participants. Consider quarterly or biannual feedback cycles for comprehensive surveys, supplemented by shorter, targeted requests as needed.
- **Simplify Surveys:** Design concise and user-friendly feedback forms that require minimal time to complete. Use clear and straightforward questions to gather essential information.
- **Communicate Purpose:** Clearly communicate the purpose and importance of each feedback request. Explain how the feedback will be used to drive improvements and benefit the respondents.

Implementation example

A retail company reduces the frequency of its customer satisfaction surveys from monthly to quarterly and simplifies the survey to five key questions. They communicate the impact of previous feedback through newsletters, demonstrating how it led to tangible improvements.

Lack of Action on Feedback

When feedback is collected but not acted upon, it can lead to disengagement and skepticism among employees and customers. People may feel that their input is ignored or undervalued.

Strategies to Overcome It:

- **Prioritize Actions:** Develop a clear process for prioritizing feedback and identifying actionable insights. Focus on high-impact areas that can deliver visible improvements.
- **Assign Responsibility:** Designate specific teams or individuals to take ownership of implementing feedback-driven changes. Ensure accountability and track progress.
- **Close the Loop:** Regularly communicate actions taken in response to feedback. Share success stories and updates to show that feedback leads to meaningful change.

Implementation example

An IT services company creates a feedback task force responsible for analyzing feedback and implementing changes. They hold monthly meetings to review progress and communicate updates to the entire organization through internal newsletters and meetings.

Insufficient Resources

Limited resources, such as time, budget, and personnel, can hinder the ability to effectively collect, analyze, and act on feedback.

Strategies to Overcome It:

- **Leverage Technology:** Use automated tools and software to streamline feedback collection and analysis. This reduces the manual workload and increases efficiency.
- **Prioritize Key Areas:** Focus resources on the most critical areas for feedback and improvement. Allocate budget and personnel strategically to maximize impact.
- **Seek External Support:** Consider outsourcing certain aspects of the feedback loop, such as data analysis or survey design, to external experts if internal resources are constrained.

Implementation example

A healthcare provider implements an automated patient feedback system that integrates with their electronic health records. This system streamlines data collection and provides real-time insights, allowing staff to focus on patient care.

Resistance to Change

Employees and management may resist changes prompted by feedback due to comfort with the status quo, fear of the unknown, or lack of understanding of the benefits.

Strategies to Overcome It:

- **Engage Stakeholders Early:** Involve employees and managers in the feedback process from the beginning. Seek their input on how to improve processes and address concerns.
- **Educate and Train:** Provide training on the importance of feedback and how to effectively use it to drive improvements. Highlight success stories and case studies to illustrate benefits.
- **Foster a Growth Mindset:** Cultivate a culture that views change as an opportunity for growth and development. Encourage experimentation and learning from both successes and failures.

Implementation example

A financial services firm holds workshops to educate employees about the benefits of feedback and the changes being implemented. They showcase examples of successful feedback-driven improvements and reward teams that embrace new processes.

Inconsistent Feedback Processes

Inconsistent feedback processes can lead to unreliable data and hinder the ability to make informed decisions. Variations in how feedback is collected, analyzed, and acted upon can create confusion and reduce effectiveness.

Strategies to Overcome It:

- **Standardize Processes:** Develop standardized procedures for feedback collection, analysis, and implementation. Ensure that all departments follow the same guidelines.
- **Document and Train:** Document feedback processes and provide training to ensure consistency. Make resources and templates available to all employees involved in the feedback loop.
- **Regular Audits:** Conduct regular audits of feedback processes to identify inconsistencies and areas for improvement. Use these audits to refine and enhance procedures.

Implementation example

A global logistics company standardizes its feedback processes by creating a central repository of feedback templates and guidelines. They conduct quarterly audits to ensure compliance and make adjustments as needed.

By identifying and addressing these common roadblocks, leaders can sustain an effective feedback loop that drives continuous improvement and long-term success. Overcoming these challenges requires a proactive approach, clear communication, and a commitment to fostering a culture that values feedback and embraces change.

Conclusion

Creating and sustaining a robust feedback loop is essential for the long-term success of an Agile Brand. A comprehensive feedback loop ensures that all aspects of the business, from marketing and customer experience to team collaboration and process efficiency, are continuously monitored and improved. By integrating both external and internal performance metrics, organizations can gain a holistic view of their operations and make informed decisions that drive sustained growth and innovation.

Leaders play a critical role in implementing and maintaining an effective feedback loop. By establishing clear objectives, fostering a culture of openness and trust, implementing standardized feedback mechanisms, and continuously analyzing and acting on feedback, leaders can ensure that the feedback loop remains a dynamic and valuable tool for the organization.

Overcoming common roadblocks, such as feedback fatigue, lack of action on feedback, insufficient resources, resistance to change, and inconsistent feedback processes, is crucial for the long-term sustainability of the feedback loop. By proactively addressing these challenges, organizations can maintain the effectiveness of their feedback mechanisms and continue to drive continuous improvement.

A well-established feedback loop is not just a tool for measuring performance; it is a cornerstone of an Agile Brand's strategy for continuous improvement, innovation, and long-term success. By valuing and leveraging feedback from all stakeholders, organizations can remain agile, responsive, and committed to delivering value to their customers and employees.

Next Best Actions

1. **Develop a Comprehensive Feedback System:**
   - Create a structured system for collecting feedback across all key areas, including marketing, customer experience, sales, product performance, and internal metrics. Ask: How can we gather and integrate feedback from all relevant sources to create a complete picture of our performance?

2. **Implement Regular Feedback Reviews:**
   - Schedule regular review sessions to analyze feedback data and identify areas for improvement. Involve cross-functional teams to ensure diverse perspectives. Ask: How can we make feedback reviews a regular part of our strategic planning and decision-making processes?

3. **Create Action Plans Based on Feedback:**

○ Develop actionable plans based on feedback insights and assign responsibilities for implementing improvements. Track progress and adjust strategies as needed. Ask: How can we ensure that feedback leads to concrete actions and measurable improvements?

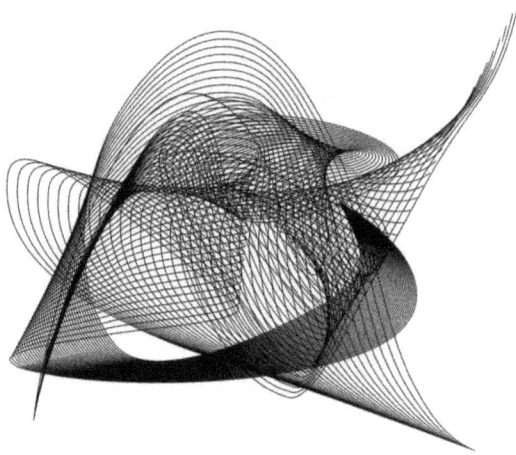

Epilogue

Every interaction in any form, is branding."
—Seth Godin

I'm so excited to have been able to share this updated version of my vision for an Agile Brand with you. So, as we reach the end of this exploration into what it takes to create an Agile Brand, let's reflect on the transformative journey that lies ahead for organizations willing to embrace these concepts. The principles of agility, continuous improvement, operational adaptivity, value-driven guidance, relationship building, conversational focus, and perpetual learning are not just theoretical constructs but practical tools that can drive profound and lasting success.

The principles outlined in this book are foundational to creating a brand that thrives in today's dynamic and fast-paced market. An Agile Brand is not defined merely by its ability to react to changes but by its proactive stance on innovation, commitment to core values, and dedication to fostering meaningful relationships with customers and employees alike.

1. **Agility by Design:** Creating systems and processes that are inherently flexible and responsive to change.
2. **Continuously Improving:** Embracing a mindset of ongoing development and enhancement.
3. **Operationalizing Adaptivity:** Embedding adaptability into the daily operations of the organization.
4. **Guided by Values:** Ensuring that all actions and decisions align with the organization's ethical standards and core principles.
5. **Building Relationships:** Focusing on long-term engagement and trust rather than short-term transactions.
6. **Focusing on the Conversation:** Engaging in meaningful, two-way communication with stakeholders.
7. **Always Learning and Growing:** Committing to continuous education and growth for both the organization and its people.

Succeeding as an Agile Brand

Success as an Agile Brand is multifaceted. It encompasses not only financial performance but also the well-being of customers and employees, as well as the sustainability of the business. When these principles are effectively implemented, organizations can achieve:

Happy Customers

By prioritizing customer satisfaction and continuously enhancing the customer experience, Agile Brands build loyalty and advocacy. The journey to achieving happy customers starts with deeply understanding their needs, preferences, and pain points. Agile Brands leverage data and feedback to gain insights into customer behavior, allowing them to tailor their products, services, and interactions to meet and exceed expectations. This customer-centric approach ensures that every touchpoint is designed to deliver value and delight, fostering a sense of loyalty and connection.

One of the key themes highlighted throughout the book is the importance of active listening and meaningful engagement with customers. By focusing on the conversation, Agile Brands engage in two-way communication, actively seeking and acting on feedback. This not only shows customers that their opinions matter but also helps the organization to stay agile and responsive to changing needs. Regularly collecting and analyzing customer feedback enables brands to make informed decisions and implement improvements that enhance the overall customer experience.

Building strong, enduring relationships with customers also involves a commitment to transparency and ethical behavior. Agile Brands guided by values ensure that all interactions are characterized by honesty and integrity. This builds trust and credibility, which are essential for long-term loyalty. Customers who feel valued and respected are more likely to become advocates for the brand, sharing their positive experiences with others and contributing to its reputation and growth.

Additionally, by continuously improving and operationalizing adaptivity, Agile Brands ensure that their customer experience evolves alongside market trends and technological advancements. This proactive approach to innovation keeps the brand relevant and competitive, ensuring that customers remain engaged and satisfied. In essence, happy customers are the result of a holistic commitment to excellence, where every aspect of the organization is aligned to create value and foster positive, lasting relationships.

Happy Employees

An Agile Brand fosters a positive and engaging work environment where employees are encouraged to innovate, collaborate, and grow. This commitment to creating a supportive and dynamic workplace begins with valuing employee input and providing opportunities for meaningful contribution. By actively seeking and incorporating employee feedback, Agile Brands demonstrate that their voices are heard and respected, which enhances overall job satisfaction and engagement. This inclusive culture empowers employees to take ownership of their work, fostering a sense of pride and commitment to the organization's success.

Encouraging innovation and collaboration is another cornerstone of an Agile Brand's approach to employee happiness. By promoting cross-functional teamwork and providing the tools and resources needed for creative problem-solving, organizations can unlock the full potential of their workforce. Employees who feel supported in their efforts to experiment and innovate are more likely to develop novel solutions and drive the company forward. This culture of continuous improvement not only benefits the organization but also contributes to the personal and professional growth of its employees.

Also, an Agile Brand recognizes the importance of professional development and lifelong learning. By offering regular training programs, mentorship opportunities, and clear pathways for career advancement, these organizations invest in their employees' futures. This focus on growth ensures that employees remain engaged and motivated as they see tangible progress in their careers. It also helps the organization stay competitive by continually enhancing the skills and capabilities of its workforce.

Lower turnover rates are a natural outcome of such a positive and engaging work environment. When employees feel valued, supported, and empowered, they are more likely to stay with the company long-term. This stability allows the organization to maintain a strong, cohesive team that can build on shared knowledge and experience. Ultimately, happy employees lead to increased productivity and a healthier organizational culture, driving the sustained success and resilience of the Agile Brand.

Successful, Sustainable Business

By aligning with core values and focusing on continuous improvement, Agile Brands achieve sustainable growth. This alignment ensures that every decision and action taken by the organization is rooted in a strong ethical foundation, fostering trust and loyalty among customers, employees, and stakeholders. Core values serve as a guiding compass, ensuring that the brand remains true to its mission and principles even as it navigates the complexities of the market. This steadfast commitment to values differentiates the brand and builds a resilient reputation that can withstand challenges and uncertainties.

Continuous improvement is at the heart of creating a successful, sustainable business. Agile Brands embrace a mindset of perpetual development, constantly seeking ways to enhance their products, services, and processes. By implementing regular feedback loops and leveraging data-driven insights, these organizations can identify areas for refinement and innovation. This proactive approach to improvement enables them to stay ahead of market trends, meet evolving customer needs, and drive operational efficiencies. It also cultivates a culture of excellence where employees are motivated to contribute to the organization's ongoing success.

Navigating market fluctuations and capitalizing on new opportunities are critical capabilities for any business aiming for long-term sustainability. Agile Brands are particularly adept at these tasks due to their inherent flexibility and responsiveness. By operationalizing adaptivity and fostering a culture of agility, these organizations can pivot quickly in response to changing conditions, seizing opportunities that others might miss. This adaptability not only helps mitigate risks but also positions the brand as a leader in innovation and customer satisfaction.

Maintaining a competitive edge in a rapidly evolving market requires more than just short-term strategies; it demands a long-term vision and commitment to sustainable practices. Agile Brands integrate sustainability into their core operations, ensuring that their growth is both economically and environmentally responsible. By focusing on sustainable business practices, they not only reduce their environmental footprint but also appeal to increasingly conscious consumers who value ethical and sustainable brands. This holistic approach to business sustainability ensures that Agile Brands are not only successful today but are also well-prepared to thrive in the future, fostering enduring success and resilience.

Where to Go from Here

The journey to becoming an Agile Brand is ongoing and requires commitment, flexibility, and a willingness to learn and adapt. Here are some steps to continue this journey:

1. **Assess Your Current State:** Conduct a thorough assessment of your organization's current alignment with the Agile Brand principles. Identify strengths and areas for improvement.
2. **Create a Roadmap:** Develop a clear roadmap for implementing Agile Brand principles. Set specific, measurable

goals and timelines and assign responsibilities to ensure accountability.

3. **Engage and Educate:** Continuously engage and educate your team about the importance of agility, innovation, and continuous improvement. Provide training and resources to help them embrace these principles in their daily work.

4. **Leverage Feedback:** Implement robust feedback loops to gather insights from customers and employees. Use this feedback to drive ongoing improvements and ensure your strategies remain aligned with stakeholder needs.

5. **Celebrate Successes:** Recognize and celebrate the achievements and milestones along the way. Highlight success stories that demonstrate the positive impact of adopting Agile Brand principles.

6. **Stay Adaptable:** Remain open to change and be prepared to pivot as needed. The market, technology, and customer preferences will continue to evolve, and maintaining agility will be key to long-term success.

Becoming an Agile Brand is a journey that requires dedication, strategic planning, and a relentless focus on continuous improvement. By embracing the principles outlined in this book, organizations can create a vibrant, dynamic, and resilient brand that delights customers, empowers employees, and achieves sustainable success.

The path forward is clear: assess where you are, define where you want to go and take deliberate steps to get there. Engage your team, leverage feedback, and remain adaptable. The rewards of this journey are immense—a thriving, Agile Brand that stands the test of time and continuously excels in a rapidly changing world.

As you move forward, remember that the principles of an Agile Brand are not just tools for today but guides for the future. Embrace them fully, and let them drive your organization to new heights of success and sustainability.

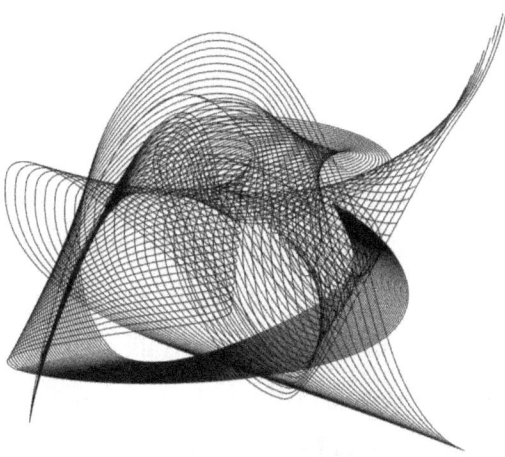

Acknowledgments

This book, both the original version I published in 2018 and this latest edition, would not exist without the support of many who have taught me so much. I can't possibly do justice to the amount of people who have been helpful, but I will attempt to, in no particular order, thank as many people as I can for the time being.

A big thanks for Chris Bach for writing the foreword to this revised edition. Thanks also to Alicia Recco for her original design for the book cover that I have used and turned into the basis for my personal brand. I'm so thankful to her for her work, and I hope she thinks it's aged well. Thanks also to previous foreword contributors to my Agile series of books, including Lisa Nirell and Bob Sprague.

Additionally, I had much help editing the words on these pages, including from my most recent editor, Loretha Green. But I'd also like to thank my sister, Janelle Kihlström-Pomery, as well as Anne-Marie Montague, for their earlier contributions.

Also, there is a very long list of people for which the knowledge and generosity they shared have formed the foundations and (for the most part) the extent of my knowledge. It would be too long a list to mention at this point, and it's grown considerably longer since 2018, when the first edition of this book was published. So suffice it to say, if I have had the privilege to work with you, talk with you, interview you, read your writing, or otherwise, a big THANK YOU for your contributions to my work and to this book.

Finally, thanks to YOU for your support. If this is the first of my books you've read, I hope you've enjoyed it and will check out others or listen to one of my podcasts, The Agile Brand with Greg Kihlström or B2B Agility. If this isn't your first, then I appreciate your continued support and hope that you are finding value in the content I've created. Please share your thoughts with me so I can create better and more valuable resources.

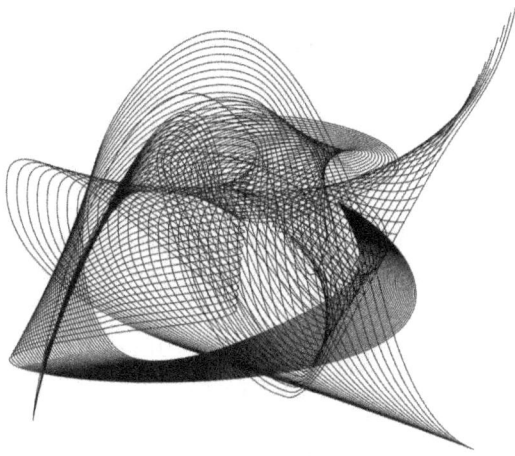

About the Author

Greg Kihlström is a best-selling author, speaker, and entrepreneur and serves as an advisor and consultant to top companies on marketing technology, marketing operations, and digital transformation initiatives. He has worked with some of the world's top brands, including Adidas, Coca-Cola, FedEx, HP, Marriott, Nationwide, Victoria's Secret, and Toyota.

He is a multiple-time Co-Founder and C-level leader, leading his digital experience agency to be acquired in 2017, successfully exited an HR technology platform provider he co-founded in 2020, and led a SaaS startup to be acquired by a leading edge computing company in 2021. He currently advises and sits on the board of a marketing technology startup.

In addition to his experience as an entrepreneur and leader, he earned his MBA, is currently a doctoral candidate for a DBA in

Business Intelligence, and teaches several courses and workshops as a member of the School of Marketing Faculty at the Association of National Advertisers. He has served on the Virginia Tech Pamplin College of Business Marketing Mentorship Advisory Board, the University of Richmond's CX Advisory Board, and was the founding Chair of the American Advertising Federation's National Innovation Committee. Greg is Lean Six Sigma Black Belt certified, is an Agile Certified Coach (ICP-ACC), and holds a certification in Business Agility (ICP-BAF).

Greg has written over 20 books on marketing and marketing technology, including his 10-part *Agile Brand Guides* series on marketing technology platforms and practices. His recent book, the best-selling *House of the Customer* (2023), discusses the 1:1 personalized customer experience of the future and how brands can organize the people, processes, and platforms that enable it.

He executive produces 5 business and marketing-related podcasts, including the award-winning *The Agile Brand with Greg Kihlström*, now top 5 on Apple's U.S. marketing charts and in its 6th year with over 500 episodes and millions of downloads, which discusses marketing technology and its role in the customer experience with some of the world's leading experts and leaders.

Greg is a contributing writer to Forbes, MarTech, CustomerThink, and CMSWire and has been featured in publications such as Advertising Age, Business Insider, Financial Times, and The Washington Post. Greg has been named #1 on its list of the Top Global Marketing Thought Leaders by Thinkers 360, was named one of ICMI's Top 25 CX Thought Leaders two years in a row, and a DC Inno 50 on

Fire as a DC trendsetter in Marketing. He's also participated as a speaker at global industry events and has guest lectured at prominent universities and colleges.

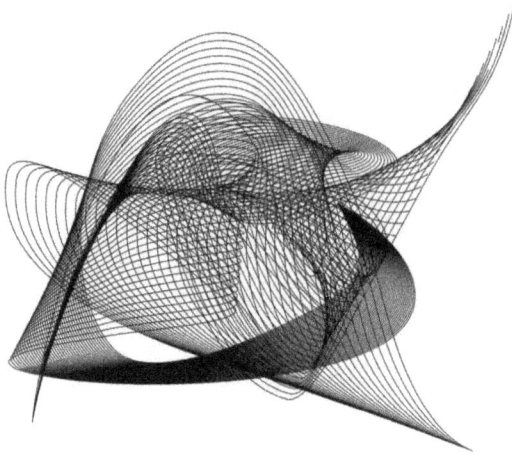

Appendix 1:
The 7 Agile Brand Principles

While they are included throughout this book, I thought it would be helpful to include just the principles here with a short description of each.

Principle 1: Agility by Design

Agility by Design means structuring your organization to be inherently flexible and responsive to change. This involves creating processes, systems, and cultures that allow for rapid adaptation to new information, market conditions, and technological advancements. By embedding agility into the design of the organization, businesses can stay ahead of competitors and quickly respond to opportunities and threats.

Principle 2: Continuously Improving

Continuous Improvement is about fostering a mindset of ongoing development and refinement. Organizations committed to this principle regularly seek out and implement incremental enhancements to processes, products, and services. This iterative approach ensures that the company always evolves and enhances its performance, leading to sustained success and innovation.

Principle 3: Operationalizing Adaptivity

Operationalizing Adaptivity involves integrating adaptability into the core operations of the organization. This means making flexibility a standard part of workflows and decision-making processes. By doing so, businesses can ensure they are prepared to pivot and adjust in response to changing market conditions, customer needs, and internal dynamics.

Principle 4: Guided by Values

Being Guided by Values means ensuring that all actions and decisions align with the organization's core principles and ethical standards. This principle emphasizes the importance of integrity, transparency, and ethical behavior in building trust with customers, employees, and other stakeholders. By adhering to their values, organizations can create a strong, consistent brand identity and foster long-term loyalty.

Principle 5: Building Relationships

Building Relationships focuses on the importance of fostering strong, long-term connections with customers, employees, and other stakeholders. This principle emphasizes engagement, trust, and mutual value creation. By prioritizing relationships over individual transactions, organizations can build loyalty and create advocates who support the brand's growth and success.

Principle 6: Focusing on the Conversation

Focusing on the Conversation means engaging in meaningful, two-way communication with stakeholders. This principle involves actively listening to feedback, facilitating open dialogue, and valuing the insights of customers and employees. By maintaining an ongoing conversation, organizations can stay attuned to stakeholder needs and preferences, driving continuous improvement and innovation.

Principle 7: Always Learning and Growing

Always Learning and Growing emphasizes the importance of continuous education and development for both the organization and its people. This principle involves fostering a culture of curiosity, encouraging innovation, and providing opportunities for learning and growth. By committing to perpetual learning, organizations can adapt to changes, drive innovation, and sustain long-term success.

These seven principles provide a comprehensive framework for organizations aiming to become Agile Brands. By integrating these principles into their strategies and operations, businesses can enhance their agility, foster continuous improvement, and build stronger relationships with their stakeholders, ultimately achieving sustained success and growth.

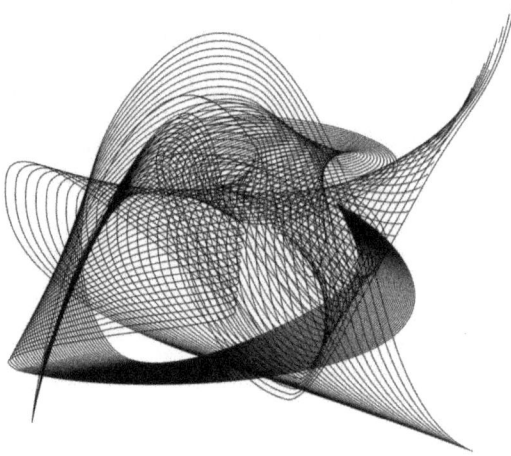

Appendix 2:
The Agile Brand Maturity Model

How Agile is your brand? We've created an Agile Brand maturity model to help guide you and your teams on their way toward becoming a more customer-focused, adaptive organization. For each of the 7 principles of an Agile Brand, we've provided three questions to ask and rate your organization on a 1-4 scale. Your score is the total number out of a total of 28.

Evaluation Scale

1 Rarely: The characteristic is rarely or never demonstrated.

2 Occasionally: The characteristic is occasionally demonstrated but not consistently.

3 Frequently: The characteristic is frequently demonstrated but still has room for improvement.

4 Always: The characteristic is consistently and effectively demonstrated.

Principle 1: Agility by Design

Characteristic	Question
Flexible Processes	How adaptable are your processes to changing market demands?
Rapid Prototyping	How quickly can your team develop and test new ideas?
Dynamic Resource Allocation	How effectively can you reallocate resources to new priorities?

Principle 2: Continuously Improving

Characteristic	Question
Regular Feedback Collection	How frequently do you collect feedback from customers and employees?
Incremental Enhancements	How often do you implement small, incremental improvements?
Performance Metrics Tracking	How consistently do you track and analyze performance metrics?

Principle 3: Operationalizing Adaptivity

Characteristic	Question
Cross-Functional Collaboration	How well do your teams collaborate across functions?
Scenario Planning	How prepared are you for potential market changes?

Adaptivity Training	How often do you provide training on adaptive methodologies?

Principle 4: Guided by Values

Characteristic	Question
Ethical Decision-Making	How consistently do you ensure decisions align with core values?
Transparency	How openly do you communicate business decisions and changes?
Value-Driven Leadership	How well do leaders exemplify the organization's values?

Principle 5: Building Relationships

Characteristic	Question
Customer Engagement	How effectively do you engage with customers beyond transactions?

| Employee Engagement | How well do you engage with employees and foster their loyalty? |
| Relationship Metrics | How regularly do you measure the quality of your relationships? |

Principle 6: Focusing on the Conversation

Characteristic	Question
Active Listening	How effectively do you listen to and act on stakeholder feedback?
Open Dialogue	How frequently do you facilitate open discussions with stakeholders?
Feedback Implementation	How well do you implement changes based on feedback received?

Principle 7: Always Learning and Growing

Characteristic	Question

Continuous Learning Programs	How regularly do you offer learning and development opportunities?
Innovation Encouragement	How effectively do you encourage and support innovation?
Knowledge Sharing	How well do you facilitate the sharing of knowledge across the organization?

Scores

To determine your maturity score, add up the totals for each question and divide by the total possible points: 84 (7 principles times 3 questions with up to 4 points each, or (7 x (3x4)). You may also choose to calculate individual principle scores, in which case each principle would have a maximum score of 12.

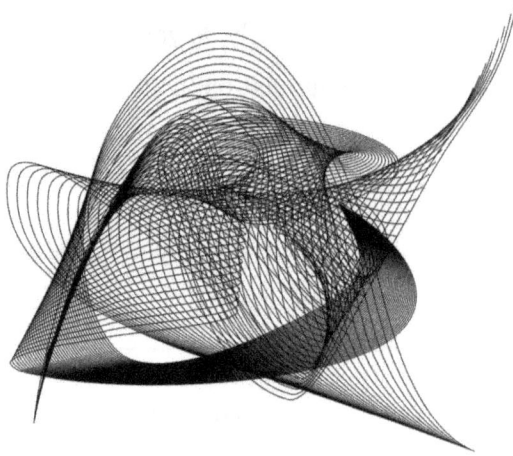

References

1 Merriam-Webster Dictionary: Agile, Information last updated September 4, 2017

2 Martin, James (1991). Rapid Application Development. Macmillan. ISBN 0-02-376775-8.

3 Kent Beck, James Grenning, Robert C. Martin, Mike Beedle, Jim Highsmith, Steve Mellor, Arie van

 Bennekum, Andrew Hunt, Ken Schwaber, Alistair Cockburn, Ron Jeffries, Jeff Sutherland, Ward Cunningham,

 Jon Kern, Dave Thomas, Martin Fowler, Brian Marick (2001). "Manifesto for Agile Software Development".

 Agile Alliance. Retrieved 14 June 2010.

4 Redding, Dan. "The History Of Logos And Logo Design." Smashing Magazine. July 6, 2010.

5 Burns, Will. End Random Acts of Marketing with a Potent Brand Idea. Forbes. May 7, 2013.

6 Bonchek, Mark and Cara France. Build Your Brand as a Relationship. Harvard Business Review. May 9, 2016.

7 Epsilon. The impact of personalization on marketing performance. January 4, 2018.

8 Forrester. "U.S. Consumers Are Willing Co-Creators." Retrieved July 7, 2024 from

 https://www.forrester.com/report/US+Consumers+Are+Willing+CoCreators/-/E-RES57506

9 Spotify. Company Info. Spotify website. Retrieve July 7, 2024 from https://newsroom.spotify.com/company-

 info/

10 Cruth, Mark. Discover the Spotify Model: What the most popular music technology company can teah us about scaling agile. Atlassian blog. Retrieved July 7, 2024 from https://www.atlassian.com/agile/agile-at-scale/spotify

11 Kniberg, Henrik and Anders Ivarsson. Scaling Agile at Spotify with Tribes, Squads, Chapters and Guilds. Retrieved July 7, 2024 from https://blog.crisp.se/wp-content/uploads/2012/11/SpotifyScaling.pdf

12 Gupta, Abhinav. What is the Spotify Model in Agile? LogRocket website. April 26, 2023. Retrieved July 7, 2024 from https://blog.logrocket.com/product-management/what-is-the-spotify-model-agile/

13 Levinson, Philip. "How Nike almost ended up with a very different name". Business Insider. Retrieved June 7, 2017.

14 Adamek, Drew. Nike's digital supply chain helps keep it running: The apparel company's investment into suppy-chain tech is a strategic advantage. CFO Brew. July 26, 2022.

15 The Guardian. How Nike Flyknit revolutionized the age-old craft of shoemaking. November 27, 2013. Retrived July 7, 2024 from: https://www.theguardian.com/sustainable-business/partner-zone-nike1

16 Cook, Grace. Nike is Grasping the Sustainability Nettle with its "Move to Zero" Campaign. British Vogue. September 30, 2020. Retrieved July 7, 2024 from https://www.vogue.co.uk/news/article/nike-move-to-zero-sustainability

17 Brettman, Alan. Nike Designer Describes Life Inside the Innovation Kitchen. Oregon Live, The Oregonian. May 11, 2011. Retrieved July 7, 2024 from: https://www.oregonlive.com/playbooksandprofits/2011/05/nike_designer_describes_life_i.html

18 https://hbr.org/2014/01/how-netflix-reinvented-hr

19 Ingraham, Nathan. Netflix plans to double its original content output next year. The Verge. May 22, 2013. Retrived July 7, 2024 from https://www.theverge.com/2013/5/22/4356864/netflix-plans-to-double-its-original-content-output-next-year

20 Whiting, Rick. Researchers Solve Netflix Challenge, Win $1 Million Prize. CRN. Retrieved July 7, 2024 from: https://www.crn.com/news/applications-os/220100498/researchers-solve-netflix-challenge-win-1-million-prize.

21 Netflix. Corporate Website. Retrieved July 7, 2024 from: https://jobs.netflix.com/culture

22 Drew, Maximus, "Netflix and Their Customer Acquisition Model" (2024). Undergraduate Theses,

Professional Papers, and Capstone Artifacts. 511. https://scholarworks.umt.edu/utpp/511

23 Keyhole. Netflix's Social Media Strategy Unveiled: A Comprehensive Analysis. Keyhold Website. Retrieved

July 7, 2024 from: https://keyhole.co/blog/netflixs-social-media-strategy/

24 Salesforce. "Salesforce Report: Nearly 90% Of Buyers Say Experience a Company Provides Matters as Much

as Products or Services." May 10, 2022. Salesforce blog.

https://www.salesforce.com/news/stories/customer-engagement-research/

25 Wagner, Daniel. "Why Corporate Values Matter, even if Not All Consumers Care." Huffington Post. February

2, 2017.

26 Smart Insights. Consumers are increasingly trusting brands that take a stance. Smart Insights website.

September 30, 2019. Retrieved July 7, 2024 from https://www.smartinsights.com/online-brand-

strategy/consumers-are-increasingly-trusting-brands-that-take-a-stance/

27 Bill Vlasic (April 20, 2003). "Toyota turns edgy to grab Gen Y buyers". Detroit News.

28 George, Patrick. "Why The Olds Bought The Second-Gen Scion xB Instead Of The Youths." Jalopnik. April 14,

2014.

29 Ewing, Stephen. "Scion iA and iM rolled into Yaris and Corolla lineups for 2017." AutoBlog. May 17, 2016.

30 Joseph, Seb. "Coca Cola centralizes social media marketing." The Drum. October 18, 2016.

www.ingramcontent.com/pod-product-compliance
Lightning Source LLC
Chambersburg PA
CBHW021942220326
41599CB00013BA/1485